Blue
Genes
and
Polyester
Plants

Also by Sharon Bertsch McGrayne

365 Surprising Scientific Facts, Breakthroughs, and Discoveries

Nobel Prize Women in Science: Their Lives, Struggles,
and Momentous Discoveries

Blue Genes and Polyester Plants

365 *More* Surprising Scientific Facts, Breakthroughs, and Discoveries

Sharon Bertsch McGrayne

Wiley Popular Science

John Wiley & Sons, Inc.

To George, Ruth Ann, Fred, and Tim

Library of Congress Cataloging-in-Publication Data
McGrayne, Sharon Bertsch.
    Blue genes and polyester plants : 365 more surprising
    scientific facts, breakthroughs, and discoveries /
Sharon Bertsch McGrayne.
      p.   cm.
    Includes bibliographical references and index.
    ISBN 0-471-14575-0 (alk. paper)
    1. Science—Miscellanea.   2. Engineering—Miscellanea.
  3. Technology—Miscellanea.   4. Discoveries in science—Miscellanea.
  5. Inventions—Miscellanea.   6. Scientific recreations—Miscellanea.
  I. Title.
  Q173.M443   1997
  500—dc20   96-33303

10 9 8 7 6 5 4 3 2

▶ Illustration Credits

Page 7, Shell Offshore, Inc.; page 9, Recognition Systems, Inc.; page 12, Cornell University; page 13, Kirt R. Williams, University of California at Berkeley; page 24, Arthur C. Aufderheide, University of Minnesota; page 34, E. P. Catts, Washington State University; page 36, Don Merton, New Zealand Department of Conservation; page 40, all, Jean Vacelet, Centre d'Océanologie de Marseille, Station Marine d'Endoume, Marseille (France); page 46, Michael Rosenberg, University of Delaware; page 48, Kenneth E. Stager, Natural History Museum of Los Angeles County; page 59, Hubert Schwable, Washington State University; page 60, H. Berkhoudt, University of Leiden (Netherlands), courtesy of Kurt Schwenk; page 63, Charles Francis; page 63, Thomas H. Kunz; page 64, Wulfila Gronenberg, University of Würzburg (Germany); page 69, Anne Savage, Roger Williams Park Zoo; page 70, Courtesy of Theodore Pietsch, University of Washington; page 71, Peter Funch, University of Copenhagen, and *Nature;* page 79, Olle Pellmyr, Vanderbilt University; page 81, Kathleen K. Treseder, Stanford University; page 82, J. W. Schopf, University of California, Los Angeles; page 83, Edward S. Ross; page 88, Donald M. Ball, Alabama Cooperative Extension Service; page 90, Robert Sommer, University of California, Davis; page 106, Peter R. Vogt, Naval Research Laboratory; page 112, David A. Greenberg, Bedford Institute of Oceanography; page 116, Scott Pitnick, Bowling Green State University; page 121, both, H. D. Bradshaw, Jr., University of Washington; page 133, AT&T Bell Laboratories; page 136, Masato Ono, Tamagawa University (Japan); page 150, Astronomical Institute of the Academy of Sciences of the Czech Republic, Ondřejov Observatory; page 157, Jet Propulsion Laboratory, Caltech; page 160, Alan P. Boss, Carnegie Institution of Washington; page 160, U.S. Dept. of the Interior, Geological Survey; page 161, Robert C. Balling, Jr.; page 167, Alexander Piel, Institute for Experimental Physics, Kiel (Germany); page 175, John Clarke, University of California at Berkeley and Lawrence Berkeley National Laboratory; page 182, all, Stanislaw P. Radziszowski, Rochester Institute of Technology, and Brendan McKay, Australian National University;

page 186, IBM U.K., Ltd., 1986; page 189, both, Gerard J. Holzmann and Björn Pehrson, *The Early History of Data Networks* (Los Alamitos, Calif.: IEEE Computer Society Press, 1995); page 191, Benno Artmann, Technical University of Darmstadt (Germany); page 193, both, Benno Artmann, Technical University of Darmstadt.

▶ Contents

▶ Preface

This book—like its predecessor, *365 Surprising Scientific Facts, Breakthroughs, and Discoveries*—aims to amuse and entertain with juicy, delectable scientific tidbits to brighten your life. From blue-jean genes to diamond churches, giant sperm, and deepest quakes, these are scientific curiosities par excellence.

Despite its frivolous purpose, this digest of scientific lore is based on recent and important new research. And for many readers, the real fun may be discovering the significance of these tall tales of science. Scientists probe and needle arcane subjects to illuminate broader problems in our world. The extremes and oddities of our universe help explain in miniature how nature works on far grander scales.

Thus the information for this book was gleaned from respected scientific journals and lay-level publications. In most cases the actual researchers themselves kindly checked the questions and answers for accuracy. The illustrations come not from professional photographers but from working scientists. Everything is documented. And suggestions for more serious reading appear at the back of the book, number-coded to correspond to the appropriate "fact, breakthrough, or discovery."

Nevertheless, the emphasis is on the fun. For the record-lovers there are the hottest rocks, the hardest substance, the smallest lightbulb, the dustiest place, the deepest bacteria, the tallest tower, the longest protein chain, and more.

For the starry-eyed, there are amazing wonders: singing dunes, cooling lasers, ticking hourglasses, dust crystals, thousand-year-old-leaves, bubble hair, gold diggers, galloping cockroaches, and flip-flopping climates. There is timely information on abortion in yucca plants and ancient Rome; how to cram for exams and how to cram fast food faster; and the optimal number of party invitations to issue. There are blue roses and blue cotton bolls; plants that grow plastic and polyester; and rivers in the sky.

Interested in something risqué? How about lactating males, the hormone of fatherly love, natural sperm banks, rough sex in the sea, and warnings for females: Semen can be dangerous to your health.

Does this still sound too light and fluffy for your tastes? For those who like their entertainment laced with self-improvement, there are opossums to teach us about aging, cotton-tops about colon cancer, megabergs about global warming, and mathematical cake-cutting about peace treaties and divorce settlements. While male birds sing for sex and territory, scientists study brain function, memory, and Alzheimer's. Raven squabbles shed light on cooperation, sperm size on sexual rivalries, avian promiscuity on marital fidelity, and so on.

For those who insist on learning more, there are suggestions for further reading about each topic. I've tried to include a range of articles at varying levels of difficulty. I've also concentrated on publications that should be readily available in a good public or college library.

National statistics paint us as science illiterates, yet a science publication exists for almost every level of expertise today. Clearly, a large segment of the population is fascinated with science and wants to know more. Among the most accessible publications are *Discover, Popular Science,* various health newsletters, *Sky & Telescope* for astronomy, and *Pacific Discovery.* The next tier up in difficulty includes some of the best and liveliest science writing today: *Science News, New Scientist, Mercury, Natural History,* and the *New York Times* science section on Tuesdays. *Scientific American,* rapidly shedding its stodgy image under its new publisher, fits somewhere between these publications and the somewhat more technical *American Scientist.* Finally, there are *Nature* and *Science:* They're filled with fabulously technical research reports, but the commentaries in the front of each issue are more broadly written and are often clear enough even for ye olde "intelligent layman." Only a few of the more specialized journals are cited because these are rarely available outside research university libraries.

The betwixt-and-between stance of the United States on metric issues is a perennial problem for science writers, of course. It's hard to please everyone. For familiar everyday sizes, I've opted for the American units we recognize instinctively: inches, miles, Fahrenheit between freezing and boiling, and so on. Meters and liters were a close call, but soft drinks have given Americans a feel for the liter, and a meter is so close to a yard that it's readily understandable, too. What about the enormous range of extreme conditions that scientists deal with routinely in their laboratories? Daily life gives us no feel for cold at absolute zero or the size of atoms, molecules, and interstellar space. So here I lapse happily back into Celsius, Kelvins, and so on. It's not a tidy way to deal with the confusion, but many American scientists juggle their professional and personal lives this way, too.

I particularly want to thank the many scientists who have given generously of their time to help make this book as accurate as possible. Like most scientists today, they want to help the public understand the light that science can shed on our lives. Among those who devoted extra time and patience to helping me understand their fields were Arthur C. Aufderheide, William Baum, Paul Beier, Michael Bevis, Steven J. Brams, Kurt Cuffey, Paul W. Ewald, Bernd Heinrich, Paul Hodges, Walter Lyons, Thomas Madsen, Demetrios Matsakis, Brendan McKay, Scott A. Miller, Nadine Nereson, Elizabeth Nesbitt, Masato Ono, Stanislaw P. Radziszowski, Michael Raff, Mark Pagel, Charles F. Raymond, Douglas C. Rees, James Rosen, Lee Rudolph, Larry Ruzzo, Samuel M. Scheiner, Hubert Schwabl, John Seidenstucker, Michael Taborsky, Alan D. Taylor, Margaret Thouless, and L. Pearce Williams. Special thanks go to family members George, Ruth Ann, and Fred Bertsch. The Seattle Aquarium, Seattle Public Library reference librarians, and the University of Washington's collections were extremely helpful, too. Obviously, where mistakes have been made, I am doubly responsible, given all the help I have received. Finally, many thanks go as always to my agent, Julian Bach, for his wise advice and guidance.

Cheers—and have fun.

Engineering

and Technology

1 "Water, water, every where
Nor any drop to drink."
—*Samuel Taylor Coleridge*

The world's driest place is home to the first new invention for providing drinking water since desalinization was patented in 1869. Where is the new drink that refreshes?

In the Atacama Desert of northern Chile, where rain falls every few *years*. There's plenty of mist opportunity, though, as heavy fog rolls in from the Pacific Ocean each night. Giant plastic nets face the fog, collecting 20 to 60 percent of the fog's moisture and between 3 and 4 liters of water per square yard of net. About 10 million fog droplets equal a large water drop, which rolls down the mesh into a trough. In the Middle East, similar nets in Oman captured even more water: about 50 liters per square yard of net.

2 **How many bytes make a movie?**

15 terabytes—10 million floppy disks—for *Snow White and the Seven Dwarfs*. Bet you didn't know there was that much informa-

tion in *Snow White*. The movie, made in 1939, was the first full-length animated feature. Its 83 minutes and 119,550 frames were digitized in 15 million million bytes for rerelease in 1993. When film is digitized, its picture is broken up into a checkerboard and each square is numbered according to the intensities of its colors. Reconstructing it by computer is the ultimate in painting-by-numbers.

 How many nuclear bombs have been detonated worldwide since Little Boy and Fat Man destroyed Hiroshima and Nagasaki in August 1945?

More than 2,036. Of these, 511 bombs exploded in air. Only 15 percent of their fallout has settled to the ground so far, but their total fallout may cause 1.2 million fatal cancers, according to some estimates. The United States conducted more tests under and above ground than any other country, while the former USSR holds the record for megatons detonated above ground. China and France resumed testing in the 1990s.

►4 **How much Velcro is needed to hang a 175-pound person on a wall?**

A. 4 square yards.
B. 4 square feet.
C. 4 square inches.

In theory, C. A 2-inch-square patch may contain 3,000 hooks and loops, each between 15-thousandths and 100-thousandths of an inch high. Yet only a third of them have to be engaged to hold up a 175-pound weight. The trick will be getting the person down. Yanking Velcro at right angles requires a force up to 20 pounds per square inch, but pulling diagonally so that each row of hooks is disconnected at a time requires only 1 to 4 pounds.

Velcro, barbed wire, and chain saws are among the few products deliberately based on natural structures. They mimic, respectively, cockleburs, Osage orange thorns, and beetle grub teeth. George de Mestral, a Swiss engineer, invented Velcro in the early

1940s after cockleburs stuck to his trousers and his dog's hair. The word Velcro combines the French words *velours* (velvet) and *crochet* (hook).

▶**5** A traveling saleswoman drives her car 200,000 miles back and forth between New York City and Detroit. A salesman drives an identical car half as many miles but makes short trips around town each day. When they put their cars on the market at the same time, which one should you buy?

Go with the woman, of course. Her car has clocked twice the mileage but made fewer starts. In general, an engine's age should be measured by the number of starts made, not miles driven. Each time the engine stops, oil drains out of its bearings. Thus when the engine is turned back on, it must operate without proper lubrication for a while until the oil pressure is restored to normal.

▶**6** Pump 10 gallons of gas into your car, and mix and match to see where it goes:

Fuel Uses	*Fuel Used*
A. Air conditioning and power steering.	1. 6 gal.
B. Tire resistance.	2. ½ gal.
C. Air drag.	3. ½ gal.
D. Braking.	4. 1¾ gal.
E. Transmission friction.	5. ½ gal.
F. Engine friction.	6. ½ gal.
G. Combustion losses.	7. ¼ gal.
	Total: 10 gals.

A–7, B–3, C–5, D–6, E–2, F–4, G–1. Only one in 10 gallons actually moves a car forward. That's the one gallon devoted to acceleration and overcoming aerodynamic resistance. Reducing the energy used for air conditioning, power steering, tires, aerodynamics, or the vehicle's mass saves six times as much energy overall. That's because those reductions decrease the burden on the engine, transmission, and fuel use. Catch a bus, anybody?

▶7 In a crisis, most drivers don't:

A. Slam on the brakes hard enough.
B. React in the nick of time.
C. Steer out of danger.
D. Remain calm; they tend to blame spouses for whatever goes wrong.

A. Seventy percent of drivers react fast enough but use only a third of the car's braking potential. A driver who applies it all can stop in half the distance. By measuring how fast the driver brakes, an electronic system could identify emergency stops and automatically apply all the car's braking power.

▶8 Most multiple-car pileups occur at speeds over 55 miles per hour, even though their drivers may be observing the famous two-second rule. Taught in driver's education, this rule requires a driver to keep two seconds of traveling time behind the vehicle ahead. So how can a driver who is following the rule cause a chain-reaction collision?

On highways drivers should use a *four*-second rule, not the two-second rule. The two-second rule assumes that the distance needed to stop a car increases directly with its speed, but the distance needed to stop a car actually varies with the *square* of its speed. Thus a driver who accelerates from 30 to 60 mph needs four times more braking distance, not just twice as much. And at 80 mph the stopping distance is more than twice what it was at 55 mph.

Furthermore, when a car decelerates, most of the slowing occurs over the *last* half of the distance. For example, if a car going 80 mph can stop in 450 feet, its speed after 225 feet is still 60 mph (not 40 mph). Consult the following table for details. And use the two-second rule for city driving and the four-second rule for highways.

Distances Needed to Stop a Car

Speed (mph)	Actual Total Stopping Distance (feet)	Two-Second Rule Distance (feet)
20	40–44	60
40	108–124	120
55	192–225	160
60	228–268	175
80	422–506	235

▶9 Using today's technology, two states could supply 60 percent of all the electricity needed by the lower 48. If Chicago is the Windy City, what are the "windy states"? And how could they electrify the rest of us?

North Dakota and South Dakota could supply 36 and 24 percent of the electricity needed, respectively. Twelve other states in the nation's midsection are breezy, too. Montana and Wyoming each could contribute 15 percent of 1990's electricity needs; Minnesota, 14; Nebraska, 7; Iowa, 5; Colorado, 9; Kansas, 12; Oklahoma, 9; and Texas, 10 percent. In fact, if areas where winds average only 14 miles per hour could be developed economically, they could supply four times more electricity than the entire United States uses now. Interest in windmills peaked during the oil crisis of the 1970s, but plummeted with world oil prices and the Reagan administration's lack of interest in renewable resource research. In the meantime, technical improvements have lowered wind power costs below those of conventional power plants.

▶10 Asphalt roads were popular in the United States during the late 1800s. It was not until 1907, however, that asphalt was mass-produced from crude oil. Where did America get the asphalt to pave its roads before that?

A lake in the Caribbean island of Trinidad off the coast of oil-rich Venezuela. The 115-acre, 80-foot-deep lake of asphalt supplied 90 percent of world demand from 1875 to 1900. Its asphalt was so hard that, when loaded bulk into ships, it fused solid and had to be chopped free at its destination. Various oil additives made it workable. It was not until 1907 that the use of asphalt manufactured from crude oil surpassed "natural lake" asphalt from Trinidad.

▶11 What's the world's tallest man-made structure? *Hint:* It is NOT the Sears Tower, which is 110 stories and 1,454 feet high.

A Shell Oil Company offshore oil rig in the Gulf of Mexico is more than twice as tall as Chicago's Sears Tower. Altogether, it rises 3,280 feet from seabed to flare top. Thirty-five "stories" are above water level. Installed in 1994, it is the world's deepest oil platform. The $1.2 billion rig is called the Auger tension leg platform, or Auger TLP for short. It's the first tension leg platform that combines both oil and gas drilling and production in United States waters. Designed to withstand 72-foot-high waves in 100-year hurricanes, it can sway up to 235 feet off center without damage. It was built to survive a 1,000-year storm.

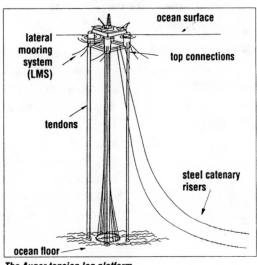

The Auger tension leg platform.

▶12 People whose stomachs get elevator jitters may want to avoid the 70-story Landmark Tower in Yokohama. Japan's tallest building boasts the world's fastest passenger elevator. How fast can it zoom?

28 miles per hour. Fortunately for the weak-livered, installing even faster elevators in even taller superhigh skyscrapers will run into technical problems.

- Doubling the airstream's speed around an elevator car increases wind noise 64 times.
- Ears hurt more the higher they rise.
- Almost 30 percent of a 100-floor skyscraper's space is lost to elevators and their engines, hoistways, and halls.
- Cables longer than about 1,300 yards cannot support their own weight. Thus new superhigh skyscrapers with more than 70 stories may have to eliminate cables altogether, substituting a ferris-wheel plan with many cars cycling continuously on magnetic rails around one shaft.

▶**13** Name the grass that supports more weight per square inch than concrete.

Bamboo. A column of the Costan Rican bamboo *Guadua angustifolia* with a surface area of one square inch can support 7,000 pounds, almost double concrete's strength. Bamboo comes in 1,250 varieties, one of which soared 4 feet in 24 hours to win a world's-fastest-growing-plant award. When bamboo houses danced through a 6.7-level earthquake in Costa Rica with nary a crack, interest in bamboo as a cheap building material spread through Latin America and Asia. Two problems remain: (1) Insects and fungi relish its starch, though salt preservatives slow them down. (2) Making weight-bearing joints from hollow tubes is difficult. Adhesives, straps, pneumatic nails, bolting plates, and wires may help.

▶**14** It identifies sperm-bank donors in Los Angeles, cafeteria diners at the University of Georgia, and outsiders in a day-care center. Kennedy and Newark airports use it to automatically check passports of international frequent fliers. In South America, Colombia's senators and members of its house of representatives use it, trying to eliminate voter fraud. What is "it"?

A. Palm reading.
B. Hand reading.
C. Fingerprints.
D Blood analysis.

B, hand reading. Place your hand palm down on a plate, and a digitizing camera takes side and top views for storage in a computer. A hand's geometry can be scanned and stored in 1.2 seconds using only 9 bytes, so few they fit on a credit card's magnetic strip. In comparison, a fingerprint may occupy 500 to 1,500 bytes.

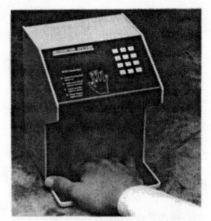

A computerized device that identifies hands.

▶15 This computerized personal identification system checks for rings and freckles, filaments, furrows, pits, striations, serpentine vasculature, and collagenous fibers. What does it look at?

The iris of an eye. The texture of an iris has six times more distinct characteristics than a fingerprint. Moreover, it remains stable for decades. Few people are likely to change it surgically; their vision would be harmed. One person can be identified in 256 bytes and recognized in 100 microseconds. Blood pressure patterns on the retina have been used too, but retina systems require subjects to place their eyes unpleasantly close to a machine.

▶16 Forget the Bronze Age and the Iron Age. The "textile age" is older than either of them. People wove textiles before they worked metal. What is the oldest known piece of weaving? Choose one of the following:

A. Fossilized linen on an antler tool handle. Turkey.
B. Cord of twisted plant fibers for weaving baskets and cloth. Israel.
C. Four quarter-sized clay pieces with weaving patterns pressed into the clay before it was fired. Czech Republic.
D. Charred cord remains. Lascaux Cave, France.

C, at 27,000 years of age. The others are (A) 9,000, (B) 19,300, and (D) 17,000 years old. Archaeologists used to think that textiles and basketry were developed after Neolithic people settled villages to raise plants and domesticate animals about 8000 B.C. Yet the

Czech weaving is so sophisticated that its makers must have been making rope and string for snares, nets, traps, and the like for centuries.

Czechs also produced the earliest known fired ceramics and the earliest known European ground stone. As temperatures cooled about 22,000 years ago, ice moved in from Scandinavia and the Alps and these talented villagers disappeared.

►17 When did people start making tools?

At least 2.6 million years ago in Ethiopia, people—or our direct ancestors—pounded roundish stones and lava pebbles into thousands of choppers, scrapers, and cutters. Archaeologists used to assume that toolmaking began about 2 million years ago with the first *Homo* species, which had a relatively large brain. But the Ethiopian tools were made about a half million years earlier, when the only known hominids alive were our even more primitive ancestors, the australopithecines. The stone tools were dated using techniques based on the natural radioactive decay of potassium and the known changes in polarity of earth's magnetic field.

►18 Where did people start making sophisticated tools, such as barbed points and blades? In Europe or Africa?

Africa. A Stone Age fishing camp in Zaire is littered with animal bone tools used to spear spawning giant catfish. Until their discovery, such sophisticated technology was thought to have originated in Europe. Zaire's technology is at least 75,000 to 90,000 years old, at least 60,000 years older than anything similar found in Europe.

►19 Gamblers handicap racehorses, but so does biology. The skeleton of a horse is fracture-prone at high speeds. How can a racehorse owner help the steed run safer *and* faster?

Shoe the horse with shock absorbers. Aluminum running shoes lined with shock- and vibration-absorbing plastic helped Thunder

Gulch win both the Kentucky Derby and the Belmont and helped Timber Country win the Preakness in 1995.

►20 **How is a silkworm like corn on the cob?**

Besides making silk, both have been domesticated so long that they could not survive in the wild. Hybrid corn, of course, depends on people to shuck its outer covering and release its seed. Silkworms (*Bombyx mori*) descended from a brown moth that lives wild in Japan, Taiwan, and eastern China, but the cultivated caterpillar and its moth have lost their camouflage and are now white. The caterpillar cannot cling to branches and doesn't even try to escape from the pan it lives in. Silkworm moths don't fly either, and some mate without ever having expanded their wings.

►21 **Who's the latest gold digger to appear in mining towns?**

Meet the metal-mining microbe, *Thiobacillus ferrooxidans*. In oxidizing inorganic materials, it releases acid and an oxidizing solution of ferric ions that washes metals out of crude ore. After chomping on copper tailings for almost 40 years, *T. ferrooxidans* accounts for 25 percent of all copper produced worldwide. Now it is pretreating gold sulfides in low-grade ores once considered worthless. The bacterium recovers more gold at less cost than traditional methods. Other microbes—including those accustomed to heavy metals or to superhot water in deep-sea vents and hot springs—may be able to biomine phosphates and other metals.

►22 **Information today races through computers a thousand times faster than in 1985. How much more information can be stored in today's computer memories?**

Only 50 times more. Moreover, the gap between computing speed and storage space is widening. Computer technology today is limited more by data storage than by computing speed. And the cost of a typical scientific workstation is determined more by the cost of its memory and disk space than by its speed.

▶**23** Scanning tunneling microscopes (STMs) were invented in the early 1980s to generate images of individual atoms and to move individual atoms around. In 1990 it took an STM the size of a thumb hours to move 35 xenon atoms until they spelled IBM. That's too big and slow if STMs are going to store vast amounts of computer data. How small is the world's smallest STM today?

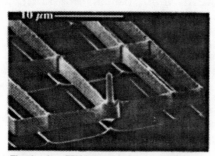

The tip of an STM is mounted on a gridlike movable platform that can be positioned precisely by electronic signals.

The diameter of a human hair. A MEM STM (that is, a microelectromechanical STM) is a microscopic atomic robot on a gold-coated silicon chip. Hundreds of thousands of them on a computer chip the size of a fingernail could help probe DNA and other molecular-scale structures, deploy automotive air bags, emulate the human ear, or hold as much data as several thousand hard disk drives today. Each bit of data would be represented by one atom or by small groups of atoms.

The tiny tip of an STM comes so close to a test sample that a current of electrons "tunnels" through the space in between. Scanning the sample's rough surface, the tip moves up and down to maintain a constant tunneling current. The record of its gyrations creates an image of the material's surface.

▶**24** What's bright enough to be seen 25 feet away in a lighted room, yet gives off 60,000 times *less* light than a 60-watt lightbulb?

The world's smallest lightbulb, a microscopic lamp with filaments as long as a human hair is wide. At their brightest, its tungsten filaments shine half a minute; at lower power they glow dim red for more than two days. Unlike silicon-based transistors, the lamp withstands high temperatures and high radiation. It could be used in a computer inside a nuclear reactor or in space near the sun. Adding several more filaments to a microlamp would turn it into a tiny vacuum tube, a microscopic version of the tubes that amplified signals in early radios and televisions. The tenth-of-a-watt microlamp converts only 2.5 percent of its energy into light, so it won't replace standard 60-watters, which are more than 20 percent efficient.

Microlamps and tubes are micromachined like computer chips. Three strips of tungsten filaments, freestanding over the rectangular cavity at center, behave like wires. A current flowing through the top filament heats that filament until it glows.

▶25 In gourmet France, military research naturally takes a gastronomic turn. Name two French culinary inventions that won both naval and army honors.

Canning and margarine. To feed sailors and explorers on long voyages, candymaker Nicholas Appert published in 1810 the first how-to canning directions: Heat food in bottles and seal the bottles. Fifty years later, Louis Pasteur explained why Appert's method worked: Heating killed the bacteria, and sealing prevented bacteria in the air from reinfecting the food. Appert invented bouillon cubes, too.

Beset by economic problems and the Franco-Prussian War, Emperor Napoleon III of France advertised for a cheap butter substitute to feed both his soldiers and the poor. In 1870 Hippolyte Mège-Mourié won with this tempting recipe: Mix milk, chopped cow's udder, beef fat, and color. No wonder the French lost the war.

▶26 The Boeing Corporation's 777 plane, debuted in 1994, is the first major commercial aircraft design that is completely "paperless." Designed and manufactured by computer, the plane involved no conventional drafting or physical mock-ups. Instead, computers linked 7,000 workstations and more than eight of IBM's largest mainframes spread across 17 time zones around the world. To help 238 project teams work simultaneously, Boeing laid high-capacity data cable across the Pacific and used a secondary satellite link for backup. The computer-aided design (CAD) software was the largest program ever installed. How much storage did it need?

3.5 terabytes. The pile of floppies would have stretched almost 5 miles high. To be precise, the program would have occupied almost 2.5 million high-density 3.5-inch disks with 1.44 megabytes on each.

▶27 Why did the crews of ancient Greek warships tie ridiculous-looking fleece pillows to their rumps?

For hindside protection. Athens's navy ruled the Mediterranean because its rowers used a sliding stroke, a technique that racing oarsmen reinvented in the mid–nineteenth century. Athenian warships fitted with battering rams were powered and maneuvered by highly trained freemen. Sitting low on their cushions, the rowers flexed their knees and then extended them straight while pulling through on the oars. A 6-inch slide adds both a foot to the oar stroke and the legs' pistonlike muscle power to the arms and back. Athens paid 12,000 rowers to train eight months a year. Their skill—and their pillows—helped build Greek culture and democracy. Carthage overtook Athens about 400 B.C. by multiplying the number of rowers on each oar, thus substituting brawn for Athens's finesse.

▶28 The world's ten largest power plants produce electricity from:

A. Nuclear fuel.
B. Water.
C. Coal.

B, water. Most of these power plants are also far from their markets. Four are in Siberia, three in remote parts of South America, and two in northern Canada. Damming a river uses its entire flow, so damming a big river generates a lot of power. Other types of power stations can produce variable amounts of power. In 1990 power plants produced enough power for everyone on earth to keep four

60-watt lightbulbs on all the time. More money is now spent distributing power than generating it. Two-thirds of the world's population are linked to a distribution network, and more than 50 countries share electrical power with a neighbor. Eventually, regional electrical distribution networks could distribute power worldwide.

▶29 | Name the world's largest exporter of electricity.

Paraguay. The world's largest power complex is the Itaipu dam on the Paraná River. Its 12,600-megawatt capacity powers Rio de Janeiro and São Paulo in Brazil.

▶30 | What Persian invention from pre-Muslim times was part of Europe's first industrial zone?

Windmills. They appeared in windswept parts of Persia before the tenth century. Greatly improved in western Europe, they were common there by 1200. In the Netherlands, 8,000 windmills ground grain and drained marshland, 900 of them at an early industrial center around a stream at Zaan. At their best, wooden windmills converted only about one-third of the wind's power to mechanical power.

▶31 | Lee de Forest, who invented the radio tube to transmit, receive, and amplify radio waves, was brought to trial in 1913 for fraud. What was the charge?

That he had used the mails to sell stock in the Radio Telephone Company. The district attorney charged that "De Forest has said in many newspapers and over his signature that it would be possible to transmit the human voice across the Atlantic before many years. Based on these absurd and deliberately misleading statements, the misguided public . . . has been persuaded to purchase stock in this company."

Absurd, eh? De Forest had been transmitting music and voice to New York City listeners since 1904 and had broadcast a live performance by Enrico Caruso at the Metropolitan Opera in 1910. Just two years after de Forest's trial, AT&T transmitted human speech from Arlington, Virginia, to Paris. Fortunately, de Forest was acquitted. His radio tubes, which made live broadcasting possible, remained the key component in radio, telephone, radar, TV, and computer systems until the development of solid-state electronics in the 1950s.

Medicine and

Health

►32 | How many Americans died from illicit drug abuse in 1990, and how many died from tobacco?

20,000 from illegal drugs and an estimated 400,000 from tobacco. The United States exports large amounts of tobacco to countries such as Colombia, yet spends millions combating the importation of cocaine, heroin, and other illegal drugs. In developing countries, cigarette smoking is increasing 3 percent a year, and total deaths caused by smoking worldwide are expected to soar from 2.5 million in 1995 to 12 million by the year 2050. In Britain half of all regular smokers will die because of tobacco use. Primarily because of smoking, an eastern European man has a greater chance of dying before age 60 than a man in India, China, or Latin America. Each cigarette costs a regular smoker 5.5 minutes of life.

►33 | What can smokers who can't stop smoking do?

According to some studies, they could get their nicotine kicks sucking snuff. If all U.S. smokers switched to snuff, tobacco-related death rates could eventually drop from an estimated 400,000 yearly to about 6,000. Smokeless tobacco such as snuff—which is sucked, not chewed—causes oral cancer, but about 75 percent of its victims survive. Only 13 percent of lung cancer victims survive.

►34 The technology already exists to (choose):

A. Halve the number of premature deaths in the United States.
B. Cut chronic disabilities by two-thirds.
C. Reduce acute disabilities by one-third.

All three. How? Exercise, wear seat belts, control guns, stop smoking and substance abuse, and eat healthfully. In short, low-cost changes in behavior and lifestyle can do more than expensive drugs or technology to improve the country's health.

►35 Assuming you are a billionaire baseball fan, should you buy a team on the West Coast or East Coast, all other things being equal?

The East Coast. When West Coast teams fly to the East Coast, they lose on average 1.24 runs to jet lag during their next game. Jet lag affects performance roughly one day for each time zone traveled. For some unknown reason, jet lag does not penalize teams flying west. In a study of 19 North American coastal teams over three seasons, 5 percent of games played by West Coast teams were affected. Those games were important. In 1991 and 1993, West Coast teams lost National League Western Division races to East Coast teams by one game.

►36 How many sperm does a man make?

Typically, at least 1,000 sperm per *minute*. That's why male contraceptives to stop sperm production hormonally have been hard to develop.

►37 Sometimes a child is born with a defect that strikes the family like a bolt from the blue. Which of the following are associated with most congenital diseases caused by a single defective gene that doesn't run in the family?

A. Older fathers.
B. Older mothers.
C. Smoking.
D. An alcoholic parent.

A, older fathers. Men—especially older men—may cause most new genetic mutations in the population. These mutations are typically associated with fathers who are an average of six years older than the fathers of unaffected children. Men produce sperm all their lives, and the sperm-making cells divide about 23 times a year. Thus by age 45, a man's sex cells have divided approximately 770 times. Each time, they can err copying a gene on a chromosome. The older the man, the more genetic mutations his sperm carry. Older fathers are associated with such birth defects as achondroplastic dwarfism, Marfan syndrome, and myositis ossificans.

Older fathers play only a slight role in Down's syndrome. Caused by an extra chromosome, it is associated primarily with older mothers.

►38 | During and immediately after both world wars, French and British soldiers fathered primarily sons. Why?

Young couples—like sex-starved soldiers—copulate more often than bored and tired-out older couples, or so the theory goes. But frequent sex clutters the cervical mucus with several hundred million dead and stranded spermatozoa, leucocytes produced by the woman, and other debris. Sperm with male chromosomes are better at penetrating thick cervical mucus.

►39 | What happens to a female gerbil who gestates in the womb between two male litter mates?

Her male neighbors in the womb bathe her in their testosterone. And as an adult, she produces primarily sons. Female gerbil fetuses with female womb neighbors produce more daughters. Men who have had Hodgkin's lymphoma or who deep-sea dive also have skewed testosterone levels. But they have lower testosterone levels than normal and produce primarily daughters.

▶40 **What do American presidents have in common with aristocrats, aggressive women, and fat possums?**

Sons. Men who have become American presidents have fathered 90 sons but only 61 daughters. Elites listed in British, German, and U.S. *Who's Who*s produce mostly sons, too. So do pregnant women judged to be assertive in psychology tests. Well-fed opossum mamas, too.

Social status explains some of the odd imbalances between sons and daughters in large populations of both animals and humans. Why is not obvious. But clearly, many birds and mammals—including deer, hamsters, rats, lemurs, horses, falcons, eagles, and several monkeys—can influence the sex of their offspring.

▶41 **Poorly fed mother opossums, spider monkeys, and South American nutria bears produce primarily daughters. Even when times are bad and weak sons can't compete successfully for mates, female offspring are sure to produce some young. Human cultures that raise more daughters, however, are extremely rare. Which of the following societies had more girls than boys? And why?**

A. Cheyenne Indians in the nineteenth century.
B. New England townsfolk in the nineteenth century.
C. The Mukogodo of Kenya.
D. The Kanjar of Pakistan.

All four, and the reasons seem to revolve around poverty. The Cheyenne were divided into upper-status "peace bands" led by peace chiefs and "war bands" led by war chiefs. The peace chiefs lived longer and their bands were wealthier than the poorer, lower-status war bands. According to an 1892 census, peace bands had about equal numbers of boys and girls, but poor war bands had about 100 girls for every 69 boys and the males died younger.

In the other cases, girls were cheaper to raise or benefited their families more. Between 1800 and 1860, urban New England families raised more girls, who earned money working in mills, while farm families raised more boys, who worked in fields. Urban girls may have received better care.

Mukogodo mothers breast-feed girls longer and get them more medical care. Men from wealthier tribes often pay a bride-price to Mukogodo families for their daughters.

Nomadic Kanjar women in Pakistan and northern India support their families by selling toys, dancing, begging, and engaging in prostitution. Kanjar men are socialized to be passive, cooperative, and subordinate to their independent and aggressive females.

Worldwide, approximately 100 million girls are missing, presumably because of abortion, infanticide, or poorer care. Boys are preferred to girls in many parts of Asia, India, and Africa.

▶42 Becoming a male is a race for time. When does the starting gun fire?

In the very beginning, when the embryo has only two cells. Male embryos develop faster so they can form testes before their mothers' estrogen levels peak. Embryos that dilly-dally develop ovaries instead of testes. This haste may explain why males reach the finishing line of life sooner. Males in 28 developed countries have shorter life expectancies than women, ranging from 3.3 years in Israel to 8.3 years in Finland.

▶43 Pregnant women who are heavy drinkers risk giving birth to children with fetal alcohol syndrome. What about drinking daddies?

They may lose their fertility altogether. After a single injection of alcohol, male rats produced half as many young as normal. Their litters were smaller, and their pups weighed less. Alcohol may have damaged their sperm—for example, by slowing its passage through the female reproductive tract. Or the alcohol could have poisoned the embryos during the first few days of development. Until recently, scientists assumed that a man's sperm was impervious to chemical damage by pesticides, alcohol, and other chemicals. Why humans abort, and the role of sperm in the abortion process, are also not well understood.

▶44 Fertility clinics often test the viability of sperm by

A. Counting the sperm.
B. Classifying their shapes.
C. Eyeballing their motility.
D. Estimating their speed.

Of these, which is the most important?

D, speed. Zippy sperm averaging 3 inches an hour are 80 percent more likely than laggards to fertilize an egg in the laboratory. Besides indicating overall vitality, speed may give sperm the momentum to penetrate the cervical mucus and the egg.

▶45 When sperm passes out of the testis, it is unable to fertilize an egg. When does sperm become fertile?

The sperm's fertility increases from a mere 5 percent to more than 75 percent as it navigates a brief, 5-millimeter-long portion of the epididymis. This convoluted passage connects the testis to the outside world, and secretions from cells lining the duct mature the sperm.

▶46 "Read! He'll never read. He can make all the letters separately; and he knows most of them separately when he sees them; . . . but he can't put them together. [He] asked me what it meant."

This description of dyslexia was written in 1853, almost 40 years before the condition was recognized. Who wrote it?

Charles Dickens in *Bleak House*. Mr. Krook, an elderly marine-store dealer in chapters 14 and 32, tries to learn to read but cannot.

▶47 Explorers and settlers brought to the New World bacteria and viruses that decimated Native American populations. Which of the following diseases were imported?

A. Chickenpox.
B. Measles.
C. Mumps.

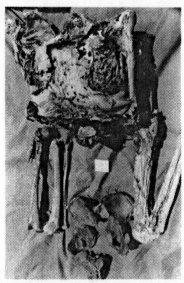

DNA unique to TB was found in a lung lesion in the body of a 40- to 45-year-old woman mummified in southern Peru 1,000 years ago.

D.   Smallpox.

E.   Tuberculosis.

A through D killed thousands of Native Americans. Tuberculosis, however, was widespread in the Americas before Columbus arrived, experts now agree. Thanks to the many tribes that collected the bones of their dead in ossuaries, epidemiological studies can be made of entire communities. TB epidemics swept Indian populations in Ontario and New York State; Mayan artists depicted hunchbacks, a deformity frequently caused by TB; and TB germs sickened a Peruvian mummy—all before Columbus.

►48 Where did tuberculosis originate in the Old World?

Among cows, as bovine tuberculosis. It probably crossed over to humans when they domesticated cattle about 8,000 years ago and then evolved into TB. TB might have developed independently in the Americas, or else it arrived with Vikings, with visitors from the South Seas, or with latecomers from across the Bering Sea. It thrives among malnourished people living in crowded quarters.

►49 When was a trip to the doctor first worth the effort?

Sometime between 1910 and 1912. Lawrence J. Henderson, an eminent physician and medical educator of the period, estimated that then "a random patient, with a random disease, consulting a doctor chosen at random had, for the first time in the history of mankind, a better than 50-50 chance of profiting from the encounter."

▶50 A male canary composes such complex songs to attract a mate and protect his territory that a large portion of his brain is devoted to musical composition. Female canaries, however, do not sing. So why do their brains have the same neuron cluster, albeit in a smaller version?

To recognize the male's signal. Otherwise, she might mate with the male of a different species. With the vocal center of her brain inactivated, she can hear the male canary singing, but she cannot distinguish his trills from the simple song of a white-crowned sparrow. One region of the brain is responsible for hearing sound, but producing and interpreting sounds must take place in another part.

▶51 Medieval artists identified shepherds by their crooks, carpenters by their tools, and wheelmakers by their wheels. What symbolized doctors and medicine?

Urine specimen bottles. Desperate for clues to what was happening inside their patients, medieval doctors scrutinized whatever clues emerged, notably urine, feces, and vomit. By the nineteenth century, medical technology had advanced somewhat and artists could afford a more highly developed sense of propriety. At that point, Laënnec's 1816 invention, the stethoscope, became the universal symbol of the healing arts.

▶52 What is the most common transplant operation in the United States?

Bone grafts. Approximately 350,000 bone grafts were performed in 1990. One bone donor can help scores of people because bone can be divided among dozens of patients or freeze-dried and stored until needed. Cornea grafts were the next most common transplant at 40,631, while skin grafts numbered 5,500. Kidneys are the most commonly transplanted organ (9,433 performed in 1990). Transplants of other organs were much less common: livers, 2,534; hearts, 1,998; pancreas/islet cells, 529; lungs, 187; and both heart and lung, 52.

▶53 In 1935, six-year-old Hildegard abandoned her embroidery to slide down the bannister with her brothers. Forgetting her sewing needle, she pricked her hand. A streptococcus infection developed into acute blood poisoning. Fourteen lancing operations later, her underarm glands were still bursting with pus and surgeons wanted to amputate her arm. Her father refused. He had discovered the world's first cure for bacterial infections. Given her father's experimental drug, Hildegard was cured. What was her father's name, and what was his drug?

Gerhard Domagk of Germany, who discovered Prontosil in 1933. Until Prontosil, the prick of a needle or rose thorn could kill. Prontosil was the first nonsurgical cure for overwhelming bacterial infections, including severe tonsillitis, abscesses, infected neck glands, puerperal fever, infections of large joints and skin, active rheumatic fever, scarlet fever, and even some fatal infections of the heart valves. Technically, Prontosil was the first effective, man-made chemotherapy for infection.

▶54 Many insect-borne pathogens thrive with split personalities. Benign to the insects that transport them around, they give their human victims dangerous diseases, such as malaria, yellow fever, typhus, and sleeping sickness. In fact, once in their human victims, the pathogens begin multiplying right off the bat so that other biting insects can pick up and pass them along to other people.

Diseases spread person-to-person are generally not as virulent. Why not?

They want their hosts well enough to travel around, infecting other people. Rhinoviruses, which cause the common cold, wouldn't get anywhere if their victims stayed sick-at-home-in-bed. But if the patients are up and about, sniffling, sneezing, and spreading their mucus hand-to-hand and hand-to-object, the pathogen travels very well indeed.

Smallpox and tuberculosis are notable exceptions. Although they are spread person-to-person, they are highly lethal. Extremely long-lived, they can manage just fine for weeks, months, or sometimes even years outside a host; hence they can afford to wait until a host comes to them. Once they find a victim, they don't care how sick their host gets because they can outsurvive any victim. Smallpox, which could live more than a decade outside a host, killed one in 10 of its victims. In comparison, a pathogen that survives only a few hours or days typically kills fewer than one in 10,000 of its victims.

►55 From traveler's trot to Montezuma's revenge, why do water-borne pathogens cause such violent diarrhea?

Grossly put, it's to their advantage. Pathogens for cholera, typhoid, and severe dysentery use water transport. They are spread from feces-soiled bed linens, clothing, and nurses to laundries and thence to drinking-water systems and more victims. Thus even violently ill, bedridden victims can spread water-borne diseases. Purified water supplies diminish both the prevalence *and* virulence of these diseases. After India began purifying its water supplies during the 1950s and 1960s, milder forms of cholera appeared. Bangladesh, where water purification lags, still suffers with the older, more dangerous form.

►56 Why is HIV less virulent in West Africa than in Central or East Africa?

Men have sex less often and with fewer partners in West Africa. The reason may be economic: In Central and East Africa, many men migrated for jobs to cities, where they tend to have many casual sexual partners. With ample opportunity to spread, the HIV virus can afford to kill its host. Much of West Africa's economy is more stable, and men have fewer sexual contacts. There HIV is less virulent: Its human host must live longer to pass the virus on to someone new.

►57 Christian Eijkman's chickens had the staggers. Swollen and paralytic, the birds swayed and twitched; some even died. Five months later in 1886, their disease mysteriously disappeared as suddenly as it had begun. Eijkman, a medical researcher for the Dutch military in colonial Indonesia, pored over his data. What had caused the disease? And what cured his chickens?

Polished rice gave the chickens beriberi, a vitamin-deficiency disease, and brown rice brought them back. The saga started when an animal keeper in Eijkman's military hospital ran out of unpolished brown rice for his chickens and quietly substituted polished white rice from the hospital kitchen. Five months later, when the hospital superintendent discovered this shocking extravagance, the birds resumed eating cheap, brown rice. Eijkman suspected that the birds had had beriberi, then known only among humans. He confirmed his suspicion by studying Javanese prison records: of 100,000 prisoners who ate brown rice, only 9 had beriberi; of 150,000 who ate polished rice, 4,000 got beriberi. Almost a half century later, scientists learned that rice husks contain thiamine, a chemical that the body needs but cannot produce itself. In the meantime, Eijkman had won a Nobel Prize.

▶**58** Name the most devastating killer-epidemic in history.

The flu epidemic of 1918. Influenza killed between 20 million and 40 million people worldwide during the winter of 1918–1919. During the month of October alone, 196,000 Americans died—more than twice the number of AIDS victims during the first decade of that epidemic. The 1918 flu struck so quickly that people sickened and died overnight. In one year, it killed as many people as Black Plague between 1347 and 1351.

▶**59** How many human diseases are known to be caused by inborn damage to a single gene?

A. 300.
B. 4,000.
C. 8,400.
D. 10,000.

B, 4,000. Diseases include muscular dystrophy, cystic fibrosis, hemophilia, sickle cell anemia, thalassemia, PKU, Hurler's Syndrome, Tay-Sachs, and SCID (severe combined immunodeficiency).

▶60 Mammography uses 40-year-old imaging technology to find small lumps in women's breasts. How many years behind the military is medical imaging technology?

A. One year.
B. Five years.
C. 10 years.
D. 20 years.

C, 10 years. Imaging technology used by the military and intelligence communities can spot tanks camouflaged behind trees and find missiles in outer space. It is just now being transferred to the early detection of breast cancer.

▶61 After 600 years of dissecting cadavers, the medical profession thought it knew every part of the human body. But a new piece was discovered in 1995. Where is it?

Between the spinal cord and the base of the skull. The half-inch-square patch of fibrous tissue is attached to a tiny neck muscle and to the dura mater membrane covering the brain and spinal cord. The dura mater is extremely sensitive, so the newly discovered tissue may play a role in tension headaches. Standard dissection techniques call for cutting through this area of the neck; two dentists and a neurosurgeon discovered the tissue because they were studying the chewing muscles and had not cut into that region.

▶62 Phineas P. Gage, a 25-year-old construction foreman for a New England railroad, suffered a ghastly dynamite accident in 1848. A yard-long iron rod blasted straight through his face, skull, and brain before landing yards away. Gage staggered from the scene with his intelligence, speech, movement, and memory intact. But in the words of his friends, "Gage was no longer Gage." How had he changed?

He had lost his moral, or ethical, judgment. Formerly widely respected and liked, the new Gage was a capricious liar who flouted social conventions, swore, broke promises, and could not keep a job. Using his preserved skull, a computerized reconstruction of the accident revealed damage to the frontal lobes of the top of his brain. Twelve other patients with brain damage to the same area also have problems making rational decisions in their personal and social lives. Thus the seat of moral behavior may be in the brain. One brain area may control knowledge of objects, language, arithmetic, and space while another operates in social and emotional domains.

►63 In 50 years what will be the three largest countries in the world? Choose among today's top-ranked countries: China, India, United States, Indonesia, Brazil, Russian Federation, and Pakistan.

India, China, and Pakistan, in that order, according to some projections. Pakistan's expected vault from seventh to third place may be caused by the low medical status of its women. The average woman in Pakistan bears 5.9 children, compared to 3.4 in India and Bangladesh. About 600 Pakistani women die for every 100,000 live births. Of every three Pakistani boys who die between the ages of one and five, five girls die; girls get less food and medical care. Worldwide, education for women is highly correlated with contraceptive use and lower birthrates. One-fifth of Pakistan's girls learn to read and write; that's half the rate of Pakistani boys or Indian girls.

India's population is expected to overtake China by the year 2045. China's birthrate has declined drastically in the past 20 years. Having reached the replacement level of 2.1 children per family, China's population could stabilize by the mid–twenty-first century. India's population, while three-quarters of China's now, is younger and has a higher birthrate.

If projections prove correct, the U.S. population will be in fourth place in 2045, followed by Nigeria, Indonesia, and Brazil. Nigeria is in tenth place now.

►64 Seattle, like much of northern Europe, abounds in cake-and-coffee houses. What's the diagnosis?

Winter depression, a form of SAD, the psychiatric condition called seasonal affective disorder. Symptoms include depression, an inability to concentrate, lethargy, a craving for sweet and starchy food, and weight gain. Eighty percent of SAD patients sleep most of the day. Problems peak during dark winters, and Seattle is located at 47 degrees north latitude. The condition may affect nearly 10 percent of the residents of New Hampshire and Alaska. Daily exposure to bright morning light for a week or two during the winter sometimes helps. The retinas of SAD sufferers may be insufficiently sensitive to low light levels.

▶**65** The colic of Poitou, the entrapado of Spain, the Huttenkatze of Germany, the bellain of Derbyshire, the dry bellyache of the Americas, and the colic of Devonshire were different names for the same epidemic. What was it?

Lead poisoning. The Romans knew that lead was poisonous when taken internally, yet they used enormous quantities to line water-carrying systems and to coat bronze and copper cooking vessels. Greek and Roman authors warned against breathing fumes from lead mines and smelting plants, drinking water near lead mines, and using lead pipes in water systems. At the same time, epidemics of lead poisoning caused epileptic-like symptoms and paralysis of the extremities, as well as colic and constipation. Ignoring public health warnings is obviously nothing new.

▶**66** Lead-laced diets can be downright lethal. What lead-rich spirits poisoned wealthy Englishmen during the eighteenth and early nineteenth centuries?

Madeira and port. The fat and ruddy-faced English gentlemen in Hogarth's prints consumed tons of Spanish, Portuguese, and Canary Island wines. These were fortified with brandy made in lead-lined apparatus. Lead prevents the kidneys from excreting uric acid. When uric acid accumulates and crystallizes in the joints, it causes the exquisite pain of gout.

More recently, Alabama moonshiners who assembled their stills from automobile radiators also suffered lead-induced gout.

▶**67** When do women excel at spatial relationships, and when do they do better at motor and verbal tasks?

During and immediately after menstruation when estrogen levels are low, most women tend to do better at spatial relationships (and worse at complex motor tasks). Just before ovulation and in the five to ten days before menstruation when estrogen levels peak, the average woman generally does better with motor and verbal work (and less well with spatial relationships). Researchers expect that male skills also rise and fall with hormonal cycles, including the 24-hour testosterone cycle.

▶**68** **What are these famous old diseases called today?**

A. Phthisis
B. Scrofula
C. Lupus vulgaris
D. Pott's disease
E. Consumption

Tuberculosis, in its many forms. A is derived from the Greek for "wasting away." B, C, and D involve neck glands, skin, and bone, respectively. TB is still the world's leading cause of death from a single infectious disease. New, drug-resistant strains have added to its virulence.

▶**69** **The killer cells that cause amoebic dysentery use the same blistering attack system that lymphocyte killer cells employ to protect the human body from infection. What is their common weapon?**

A hole-boring protein. Simply put, the protein shoots enemy cells so full of holes that they leak and die. The parasitic amoeba responsible for severe dysentery contains a potent protein that attacks and kills cells within the intestine of its unfortunate victim. After binding to a target cell, the amoeba secretes a lethal protein that forms large tubular pores in the cell membrane. Covered with holes, the target cell swells with water, blisters, and dies.

Killer cells in the human immune system attack tumor and virus-infected cells the same way. The technique may be involved in other forms of cell death as well.

Ecology and
Animal Behavior

70 One fine summer's day, the nude beach at Point Reyes, California, was filled with young people hoping to meet and possibly mate. Meanwhile, just up a hillside path, a group of male bot flies also collected in hopes of meeting and mating with females. Soon scientists invaded the scene to have a good look—at the flies, of course. Why were they more interested in bots than bare-breasted, bare-bottomed bods?

A woodrat bot fly waiting for a virgin fly.

Because adult bot flies—especially adult *male* bot flies—are far rarer than nudes at Point Reyes. Bot flies spend most of their lives as maggots inside mammals. There are nose and throat bots; rodent and rabbit bots; heel flies and stomach bots. Nose bots, for example, over-winter as maggots inside the throats of deer, elk, and moose. To clear his stuffy nose of grown-up, 3-inch-long maggots, a moose simply sneezes them out. As flying adults, bachelor bots have only a few days to attract a mate. To facilitate the process, males collect in a *lek,* an insect singles bar of sorts. Waiting for virgin females to wander by and pick them up, the males stake out territories within the lek. Fly territoriality was discovered in bot fly leks.

As an example of this charming process, male horse stomach bots spend one to three days hovering 30 minutes at a stretch directly behind the front legs of a horse as they wait for a virgin fly to fly by. Bots mate and lay eggs on the horse's forelegs. Grooming horses swallow the eggs. In winter, the eggs hatch and maggots grow within ulcers on the horse's stomach lining. In spring or summer, the maggots escape in the horse's feces. After a short time as pupae, flies emerge to hover, lek, and mate once more.

Is it any wonder those entomologists preferred bots to bods?

►71 Ever since the Renaissance, Western civilization has viewed ancient Greece as an Arcadia: a mythical place where happy human beings live in innocent harmony with the land. If that's true, why is much of Greece a barren, stony wasteland?

Eight thousand years of deforestation and soil erosion. It began as early as 4000 B.C. and has continued through virtually every historical era to the present. Today archaeologists make cultural and natural histories of entire regions, rather than just single sites. They analyze pollen and soil, use remote sensing (including ground-penetrating radar and satellite images), and record every artifact, site, and feature. Their conclusion? During peak growth periods, Greek farmers and grazers moved up mountain slopes and hillsides, clearing mixed forests that stabilized the soil. Catastrophic erosion carried soil to river bottoms. When population declined, soils were only partially rebuilt. Apparently, the ancients were no more careful of nature than we moderns are. So much for Arcadia.

►72 What country has 0.2 percent of the world's landmass but 11 percent of the world's endangered birds?

New Zealand, the aviary of the South Seas. In 1994, 98 of its bird species were listed as rare, endangered, or vulnerable. Separated from other landmasses, New Zealand developed free of predatory mammals. Many of its birds grew enormous, nested under or on the ground, or were unable to fly or glide. British settlers formed "acclimatization societies" to rectify the situation by importing proper British mammals: rabbits, pigs, deer, weasels, ferrets, stoats, and more. Rats and mice came along for the ride. The effect on New Zealand's birds and animals was catastrophic. Even local plants, unused to being grazed by herbivores, were devastated.

►**73** When Old Blue died, a formal proclamation of the New Zealand parliament announced her passing as if the old lady had been royalty. Who was Old Blue and what had she done to deserve the honor?

"Bridget," a juvenile black robin on South East Island, Chatham Islands, New Zealand. December 1988.

Old Blue—a Chatham Island black robin the size of a sparrow—rescued her species from extinction. In 1976, seven remaining black robins (*Petroica traversi* Buller) were the most highly endangered birds in the world. They lived in a dying scrub forest on a rock island 500 miles east of New Zealand. To save them, biologist Don Merton cleared a nearby island of wild sheep. Then he moved every black robin on earth to the new home.

By 1980, five black robins remained, including Old Blue. The only breeding female extant, she was already two years older than the species' life expectancy. Old Blue rose to the occasion. And with Merton transferring her eggs to tomtit nests on another island, she produced enough young to raise the population of black robins to more than 30. By the time she died in 1984, she was at least 13.

The technique of clearing islands as safe havens for highly endangered species has been adopted elsewhere, and the population of the black robins is now more than 200—and growing.

►**74** A reasonably industrious squirrel can hide thousands of *half-eaten* acorns for winter. Its recovery rate is something else again. In a good acorn year, squirrels may fail to find almost _____ percent of the acorns they hid just a few months before. Choose from the following:

A. 25 percent.
B. 50 percent.
C. 75 percent.

C, 74 percent, to be precise. Why so few? It could be forgetfulness or simply squirrel mortality. As for those half-eaten acorns, squirrels (and blue jays and grackles) chomp off the acorn's top, where weevils often grow. Then the squirrel hides the acorn under leaf litter— on the off chance he can locate it in winter. Many of these half-eaten acorns generate to start second-growth oak forests.

▶75 During the winter of 1858, a group of soldiers in southwestern Wyoming ran out of food and marched 12 days back to Santa Fe, New Mexico, for provisions. By slaughtering their starving pack animals, each man could eat 5 or 6 pounds of meat daily. But instead of thriving on the high-protein diet, they became so thin and weak they barely survived. They needed some kind of Hamburger Helper to supplement the protein. What was it?

Fat. For today's fat-phobic eater, lean is good and fat is mean. But the meat of starving animals is almost fat-free and can cause protein poisoning. Carnivores in feast-or-famine habitats, for example, often refuse to eat fat-depleted meat; one spring day around A.D. 1450, humans hunting in New Mexico butchered bisons and ate the males, but abandoned the fat-depleted females. A starving mammal's last fat reservoir lies in its bone marrow, and bone-crushing mammals have evolved (and disappeared) at least six times during the past 65 million years. Strong jaws enabled them to survive during lean—literally lean, that is—periods. The spotted hyena of Africa is a living example of a bone-crushing carnivore in a feast-or-famine habitat.

▶76 In 1856, the Reverend F. O. Morris praised the little dunnock bird for its humility as if it were a member of the "lower classes" and knew its proper place in life: "Unobtrusive, quiet and retiring, without being shy, humble and homely in its deportment and habits, sober and unpretending in its dress, while still neat and graceful, the dunnock exhibits a pattern which many of a higher grade might imitate, with advantage to themselves and benefit to others through an improved example" (*A History of British Birds*). Ac-

tually, the Reverend Morris would have been horrified by the truth about the dunnock bird. How so?

The dunnock enjoys an extraordinarily complicated sex life. It includes both monogamy and polygamy of various sorts, all embellished with an extraordinary amount of flagrant promiscuity.

In brief, females stake out territories and defend them from other females. Males offer to help out. Sometimes a single male sticks with one female (monogamy) or two adjacent females (polygyny). Or he and another unrelated male may share one female (polyandry) or a harem of *several* adjacent females (polygynandry). The dominant male tries to guard the female as closely as possible, but she keeps slipping away and soliciting sex with the subordinate male. By sharing her charms, she gets both males to help feed her chicks. DNA fingerprinting discovered the dunnock's lurid secrets.

▶**77** A female dunnock bird with two male mates may copulate six times an hour, or several hundred times per clutch. What does she do with the sperm she collects?

She pools it in roughly 1,400 tubes between her uterus and vagina, like many other breeding female birds. When the female dunnock solicits a male, he pecks at her cloacal opening for a minute—rather like a customer tapping at a speakeasy. When she ejects a droplet of sperm collected during her previous wanderings, the male hops aboard to copulate. In the case of two males and one female, each male helps feed the chicks, but only in proportion to the number of copulations the female granted him. He apparently keeps score.

Monogamous birds typically copulate much less often, between one and 20 times per clutch. But the dunnock's sex life, while a trifle florid, is not unique. Extra-pair matings are common among seemingly monogamous birds. That we have also learned from DNA fingerprinting.

▶**78** Most temperate zone mammals give birth in the spring when food becomes plentiful again, yet one large mammal comes into estrus year-round and bears litters any month of the year. Even in the north, one-quarter of its litters may be born on the ground in open air between October

and March. What is this hardy animal? And why this cold weather mother-hood?

The puma, *Felis concolor,* aka cougar, mountain lion, panther, painter, and catamount. Their main food sources, deer and elk, are more accessible in winter than in summer. In winter deer and elk move below the snow line, bunch up, and may be weakened by snow or starvation. Competing predators—black bears and griz-zlies—are hibernating.

Radio collars have enabled scientists to learn about pumas, often called the largest purring cat.

▶79   **What's the land mammal with the biggest range in the Western Hemi-sphere? It once flourished from the Yukon to the Strait of Magellan and from the Atlantic to the Pacific.**

Pumas again. Subspecies stalk high mountains, Pacific rain forests, southwestern deserts, Florida Everglades swamps, and grasslands. By World War I, pumas had been driven almost to extinction in North America by widespread poisoning, bounty systems, government-paid hunters, and the overhunting of hoofed mammals, their chief prey. Now, thanks to big populations of deer and elk and bans on poisoning, pumas are common again throughout much of the western United States.

▶80   **Name the predator that catches prey seven or eight times larger than itself.**

Pumas yet again. A 100-pound female has been seen killing an 800-pound bull elk. As puma populations have rebounded and human populations have exploded in the western United States, so has the number of puma attacks on people. Subdivisions in western canyons invade prime habitat for female pumas with young, and lawns and shrubs attract deer for pumas. Since 1890, approximately 60 puma attacks have occurred, 43 of them since 1970, at least 7 of them fatal. Juveniles are believed responsible for most attacks on humans.

On meeting a puma in the wild, try to convince it that you are not prey. Therefore, do not run, crouch, or bend over. Try to look large by opening your jacket; make eye contact; wave your arms slowly; and speak loudly and firmly. Throw anything you can reach without bending down or turning around. Above all, don't run, because that arouses the cat's instinct to chase.

▶ **81** Sponges are filter feeders, par excellence. They sit like quiet passive lumps, pumping water in and out, sieving tiny particles for food. How does a hungry sponge survive in still and food-poor water?

A. A hungry sponge waits alluringly for dinner. *B. Five minutes after it has captured a tiny mysid shrimp.* *C. Twenty-two hours after capture, the shrimp is engulfed and the sponge is digesting. Magnification: 1.6*

By becoming carnivorous and luring small creatures to their deaths. Ditching its pumping and filtering apparatus, the Mediterranean sponge *Asbestopluma* grows Velcro-like filaments to trap tiny crustaceans. Victims wiggle for hours as the filaments grow and engulf them. Within days, the sponge digests its catch. Scientists think the meat-eater evolved to cope with the scarcity of food in deep water.

▶ **82** Boom-and-bust life cycles aren't limited to Las Vegas gamblers, oil prospectors, and Wall Street traders. Some land mammals also specialize in roller-coaster modes of operation. Their populations explode and die off to 10 to 30 percent of their peak numbers at regular intervals. Which of the following animals have grown accustomed to this lifestyle?

A. Corsican mouflon sheep.
B. Lemmings.
C. Mice.

D. Muskrats.
E. Snowshoe hares.
F. Sheep.
G. Voles.

All of them, A through G. Until recently, zoologists thought that all-or-nothing population cycles were limited to small arctic and sub-arctic mammals. But large mammals are prone to life-and-death excesses, too. Feral sheep, which have populated Soay Island off Scotland for more than 1,000 years, have skeletons like early Neolithic sheep. A flock on a neighboring island peaks at about 1,400 every third or fourth year, after which roughly 70 percent of the sheep starve to death. Without this periodic die-off, the entire flock would perish. A year later, its population has rebounded; in fact, it doubles. The ewes breed before their first birthday, the island lacks predators, and plants grow fast in northern summers.

The length of a species' boom-to-bust cycle increases with the animals' body size. Corsican mouflon sheep, introduced to the sub-arctic Kerguelen Islands in the 1950s, cycle every four years.

▶83 Emperor penguins are so large—3 feet tall and 60 pounds, typically—that their young cannot mature during a brief Antarctic summer. What's a poor papa penguin to do?

Form a nursery and stand tight. The female emperor (*Aptenodytes forsteri*) lays her egg during the dead of winter and immediately leaves to deep-sea dive for food. The male emperor is left standing without food for nine weeks, cradling the egg between his feet, a fold of abdominal skin that resembles a pouch, and his warm, bare belly. Talk about being stood up. To keep warm in the face of high winds, blizzards, −40°F temperatures, and 24-hour darkness, thousands of male emperors huddle together. On the outside of the crowd only a short time, they work their way toward the center to warm up. A scientist flying over Antarctica recently saw 7,000 male emperors huddling on exposed sea ice far from shelter.

Month-old emperor chicks also cuddle for warmth. Grouping by the hundreds reduces an individual's heat loss by an estimated 25 to 50 percent. In fact, chick crèches generate so much heat that the ice melts under them. The babies must change locations often. Is it any wonder that only a few species inhabit Antartica proper?

▶84 The dominant plant-eater of the New World's tropics harvests roughly 20 percent of the fresh leaves there. It virtually controls the ecosystems of many regions. Who's this chief of the leaf-eaters?

The leaf-cutting ant. In a major evolutionary breakthrough 50 million years ago, attine ants began harvesting leaves to grow fungi for food. Today leaf-cutters are the most advanced of 200 fungus-growing ants in Central and South America. A nest of 8 million leaf-cutters can strip a large tree bare in days. After moving a living room's worth of soil to build underground chambers, the ants chop leaves into bits to manure the fungi. Every leaf-cutting farm descends from the same ancestral fungus: for at least 25 million years, leaf-cutter queens have started new farms with bites from the old homestead. To compete, Dannon would have to maintain the same yogurt culture for 25 million years. Leaf-cutting ants are the earthworms of the tropics, recycling plants and aerating soils. Thanks to their leaf-chomping prowess, however, it's hard to farm in the New World tropics.

Fungi may be the basis for all terrestrial ecosystems and there may be six times more fungi than plants in the world, but only a few scientists study molds.

▶85 Name the southern queen that is 70 percent ovary and can lay up to 5,000 eggs a day.

A red fire ant queen, *Solenopsis invicta*. Within 60 years, the tiny red Brazilian (or Argentinian) import has colonized most of the southeastern United States. Whether here or in South America, most fire ants live in colonies with 100,000 to 200,000 workers and *one* queen. But some *S. invictas* live in "supercolonies" composed of many mounds, each containing 100,000 ants and scores of egg-laying queens. Up to 300 such mounds interconnected by tunnels can blanket one acre. Outcompeting other insects, *S. invictas* attack en masse, stinging repeatedly and painfully. The multiple-queen type of *S. invicta* was discovered during the 1970s and is spreading from the Southeast toward California.

During the 1960s, World War II bombers doused entire counties with poison ant bait. The effect: By killing competitors, the poison helped fire ants spread and dominate native ants. Importing *S. invicta's* enemy insects and diseases from South America could help postpone a red ant alert.

▶86 Ecologically speaking, the most important animal in Australia may be the _____. Fill in the blank with one of the following.

A. Ant.
B. Kangaroo.
C. Dung beetle.
D. Burrowing wombat.
E. Marsupial mole.
F. Marsupial mouse.

A, ants, some Australian biologists say. The combined mass of Australia's ants is bigger than the mass of all its vertebrates. Within a few acres of semiarid Australia, 150 ant species may live, spreading seed, aerating soil, and feeding lizards and other insectivores. Because ant colonies respond quickly to environmental disruptions, they may be ideal bioindicators of a habitat's health.

▶87 What place has the highest density of predators on earth? In other words, what's a good spot to avoid?

The floodplains of the Adelaide River in Darwin, Australia. They crawl with snakes and rats. One thousand water pythons (*Liasis fuscus*) can inhabit one square mile of the floodplain. For those planning to visit the site, that's one ton of predatory pythons built to constrict their prey to death.

Area rainfall is erratic to say the least, varying between 30 and 80 inches yearly. In a medium rainfall year, those pythons share that square mile with more than a quarter million dusky rats (*Rattus colletti*). If you're still with me, that's 20 more tons of slithering biomass per square mile. During years that are too wet or too dry, the population of rats—the pythons' main food—drops dramatically to a mere 40,000, or 2.5 tons per square mile.

Pythons are uniquely able to cope through thick and thin, as it were. In good years they eat rats, breed, and make merry, and almost all their young survive. In bad years, they shut down. Females stop ovulating, and both males and females virtually stop eating. Pythons can remain quite healthy without food for up to a year. Because they are cold-blooded, they do not have to eat to maintain a high body temperature; the sun does that. Thus they survive on stored fat and muscle.

Warm-blooded predators can't compete, because they need a constant food supply. So pythons can savor the rats without fear of competition *or* attack.

▶**88** This is one cool customer. In 110°F weather, a 200-pound male gets by drinking two or three quarts of water every week or two. Most of the year, it gets all its water from food, retaining moisture by excreting the salts from its urine. It dissipates heat with a minimum of water loss. By wiping its drippy nose and drooling mouth on its forearms, it cools the fine blood vessels under its fur. Its fluid-filled foregut—10 percent of its body weight—doubles as an emergency water supply. Name this super-cool critter.

The red kangaroo. Ironically, it was once considered a "primitive" mammal with a poorly developed thermoregulatory system. Australia's plains are normally so dry that few red kangaroos survive weaning. Those lucky enough to be born during a series of wet years may live 30 years or more, however. Both territorial and sedentary, red kangaroos spend most of their time resting. They use less energy hopping 15 or 25 miles per hour than "walking." Females weigh only 50 to 65 pounds.

▶**89** What region has lost more mammals in modern times than any other place on earth? And why?

The arid outback of central Australia, made famous by the book *A Town Like Alice*. In 200 years of settlement the outback has earned an unenviable world record by losing more than half its mammal species—chiefly through overgrazing. Central Australia's outback can go years between rainstorms, so most of its mammals are nocturnal and burrow underground to escape daytime heat. The soil is among the world's thinnest, and both ground and vegetation are accustomed to soft-footed kangaroos, not hard-hooved ruminants. Yet until recently, ranchers were encouraged by "drought" subsidies to overstock the region.

Most of the extinct and endangered creatures are medium-sized, for example, the extinct stick-nest rat and the endangered numbat and bilby. Medium-sized animals need more nutritious vegetation than reptiles and small mammals but cannot migrate long distances for water, as larger herbivores can. Satellite-based soil surveys could distinguish temporary vegetation losses caused by climate from more permanent losses caused by overgrazing and trampled soils.

▶ **90** Houseflies infected with the fungus *Entomophthora muscae* die— but not before acquiring a fatal attraction. Without it, the fungus would die with the fly. What ensures that the fungus spores will live to infect other flies?

The fungus makes the fly look pregnant and chock-full of eggs. Male flies attracted to the corpse's swollen abdomen mate with the body, become infected, and die too—but not before they've passed the infection on to another living female. It's a sexually transmitted disease, fly-style.

Given the choice between a dead fly infected with the fungus and a normal dead fly, males opt for the big-bellied body. Apparently fertile females are especially alluring. The fungus may also throw in an enticing scent to make sure the males fall for the bait.

▶ **91** A mysterious pestilence has wiped out all the forest birds on Guam. Hawaii's already hard-pressed birds may be next. Name the culprit that killed Guam's birds.

A. Pesticides.
B. Avian diseases from domesticated chickens and pigeons.
C. Birds introduced from the mainland that compete with or harass native species.

D. Brown tree snakes.

E. Rats, cats, and other mammals.

D, the brown tree snake *Boiga irregularis*. A, B, C, and E were suspects in the case until a researcher thought to bait forest traps with small chickens. When she returned a day later, the chickens were gone; in their place were fat and happy *Boiga*s. The tree snakes came to Guam just after World War II by accident, probably as stowaways on military transports from the Solomon Islands. By 1986 Guam's forests were silent and the Guam flycatcher was extinct.

Remote islands are particularly vulnerable to introduced predators because they have fewer native species than areas the same size on a continent. Hawaii, the most remote archipelago on earth, has no native snakes.

Incidentally, *Boiga*s are venomous: in 1993 their bites hospitalized 50 people on Guam.

▶**92** **Which did people domesticate for food first—pigs or plants?**

An excavated building in the village of Hallan Cemi, Turkey, where pigs were first domesticated.

Pigs. Ruins of a small village inhabited 10,000 to 10,400 years ago in southeastern Turkey included piles of pig bones, but no sign of wheat or barley. Wolves were domesticated earlier as "man's best friend," not as food. Sheep and goats were tamed and herded about 1,000 years later.

Inhabitants of the tiny stone village in the Taurus Mountains ate primarily male pigs, many less than a year old; apparently they saved females for breeding. The pigs' molars were smaller than those of wild boars, indicating domestication. Pigs are easily caught and tamed, and they forage for themselves. They convert 35 percent of their food into meat, more efficiently than sheep at 13 percent or cattle at 6.5 percent.

Archaeologists relying on data from farther south in Syria and the Jordan River Valley had surmised that people there settled down first to farm wild grains. The conclusion? Perhaps people stopped roaming for different reasons in different places.

▶**93** Funk Island, a desolate granite rock off Newfoundland, was the New World's first pit stop for fast-food take-outs. Millions of great auks lived around the North Atlantic in prehistoric times, and 200,000 nested on Funk Island in the 1500s. Sailors stopped there for 400 years, provisioning their ships with the goose-sized, flightless birds. Superb underwater swimmers and fish hunters, great auks were docile and defenseless on land. Today the great auk is extinct; the last three were strangled on June 2, 1844. What drove great auks to extinction?

A. Meat-eating sailors.
B. Feather beds and pillows.
C. Scientists.
D. Its inability to fly.
E. Its habit of breeding in large numbers in a few places.

B, C, and E. Crews lived on Funk Island each summer from 1750 to 1800 gathering tons of feathers. Great auks (*Pinguinus impennis*) were rich in fat and Funk Island had no firewood, so thousands of the birds were burned as fuel to boil pots of water; then more birds were thrown alive into the pots to loosen their feathers. Tens of thousands of composted auks form Funk Island's only soil. By 1800 scientists knew the great auk was in trouble, but continued killing the birds for museums and collectors. In the 14 years before great auks became extinct, 60 were killed for their skins just from colonies off southwestern Iceland. The great auk's habit of congregating in only a few places made it an easy catch.

▶**94** Wherever African wild dogs are common, their chief prey is scarce. Are these dogs stupid hunters or what?

No, they're just avoiding lions, which prey on African wild dog pups. African wild dogs are fierce predators—they disembowel antelopes—but they would rather live where food is scarce than let their pups meet a lion.

The African wild dog's scientific name, *Lycaon pictus*, means "painted wolf," but it's not closely related to dogs, foxes, wolves, or jackals. German shepherd–sized, they travel in packs with eight or ten close relatives over large areas of 100 to 1,000 square miles. Once found in woodlands, bushlands, and plains from Senegal to South Africa, they are rapidly approaching extinction. Farmers and wildlife managers shot thousands as vermin during the 1960s and 1970s. Only 5,000 to 6,000 remain in Africa, 3,000 of them in protected areas. Thus African wild dogs are as threatened as black and white rhinos. African wild dogs are also plagued by disease and motor vehicles; in some areas the latter kill 12 percent of the adults and 20 percent of the puppies yearly.

► **95** Given the fact that only one animal dies each day in an average square mile of healthy forest, how do turkey vultures beat insects and mammals to the putrid flesh?

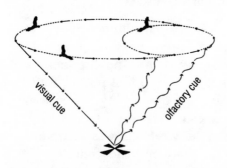

Once a turkey vulture smells a fresh carcass, it circles until it can see the animal.

Turkey vultures can smell rotten meat from afar. The scent regions of their brains are highly developed. Soaring and gliding for hours without flapping their broad wings, turkey vultures need little energy to conduct an aromatic survey of a forest. Ironically, the vultures with the glamorous reputations—condors and black vultures—don't search for food. They simply fly high enough to spot any excited turkey vultures that have sniffed out meat. The system works: When researchers set out carcasses, vultures found 60 to 95 percent of them. In a tropical forest, vultures eat more meat than all the mammals put together.

Incidentally, vultures prefer slightly aged flesh. Carrion must be

12 or more hours old before a turkey vulture can smell it, but only the hungriest bird will eat a truly rotten carcass.

▶**96** | **What is a medieval French/Chinese/Guadeloupean pig doing on Haiti's peasant farms?**

Replacing more than a million native Haitian pigs eradicated during the early 1980s by Jean-Claude "Baby Doc" Duvalier. Haitian pigs descended from medieval European swine brought to the Caribbean by pirates during the seventeenth and eighteenth centuries. Small and wiry, local pigs thrived on scraps and mangoes, producing piglets for food and sale. When the incurable African swine fever broke out in neighboring Hispaniola, the U.S. government offered $15 to $40 bounties for dead Haitian pigs. Duvalier's soldiers, lured by the prospect of wealth, conducted search-and-destroy missions. With all the pigs gone, peasants chopped down their mango trees and bought goats, in two blows worsening Haiti's erosion problem. French scientists "re-created" the extinct Haitian pig by crossing tough, prolific Chinese pigs with Caribbean and medieval Gascon breeds. When the Duvalier regime fell, the French sent their creation to Haiti, where it now accounts for more than a third of all porkers.

▶**97** | **Fast food is great for people on the run. But birds on the fly thrive on fast food, too. Sometimes they chow down even faster than their digestions work. How do these small creatures pack it in and digest it fast, without reaching for the Pepto-Bismol?**

They grow more gut—up to 22 percent more. They need every inch of it, too. Catching an insect every three minutes, a house wren collects food twice as fast as it can digest a meal. In a one-day fuelstop, a migrating songbird can gain 10 percent of its body weight. Shorebirds such as sanderlings can raise their body weight 40 percent in two or three weeks.

To cope, other birds increase digestive efficiency. When robins switch from summer insects to winter fruits, much of what goes in comes out in 40 minutes flat. And robins are five to eight times better at absorbing fruit sugar across their intestinal walls than insectivorous house wrens.

▶**98** In Israel's Negev, a desert where fewer than 4 inches of rain fall each year, lichens live *inside* limestone rocks. They flourish just *under* the rock surface, in between rock particles. There they get sunlight for photosynthesis and moisture from dew. Oddly, most of the rocks are etched 0.5 to 2.0 millimeters deep with a lacy filigree of white lines.

Is something attacking the lichens? Will the rocks survive the "Attack of the Lacy Lines"?

Yes to the first question, and no to the second. Small nocturnal snails file away the limestone to eat both rock and lichens. Their file is a tongue-like, toothed organ that grows continuously and replaces itself as it wears out.

Three species of half-inch-long snails form most of the Negev's new soil. Digesting only 5 percent of what they eat and eliminating the rest, the 100,000 snails that live on a typical acre can convert 800 pounds of rock into soil annually. In comparison, winds from the Arabian and Sinai Peninsulas blow in only 220 to 420 pounds of dirt. The most common snail species also contributes 11 percent of the nitrogen entering Negev's soil. During summer's heat, the snails stop feeding and plug their shells closed until fall. And we thought we kept our noses to the grindstone!

▶**99** What creature set the record for suspended animation? *Hint:* Its embryos hatched after being buried 10,000 years in deep sediment near Great Salt Lake in Utah.

It's a brine shrimp (*Artemia*), a crustacean related to water fleas, not shrimp. Thanks to its nearly impenetrable shell and a pump to remove excess salt, a brine shrimp can survive saltier water than any other multicellular organism. When the salt in their pond or lake water is about to crystallize, many brine shrimp stop producing live offspring and start encasing their embryos in cysts. Desiccated cysts are sold as fish food in pet stores or as "sea monkeys" that "hatch before your very eyes." A few cysts unearthed by oil drillers near Great Salt Lake hatched when placed in water. They had survived 10,000 years as static, nonliving cells and did not consume oxygen at any rate detectable by sensitive instruments today.

▶100 What's the oldest creature known to have given up sex?

A Mediterranean brine shrimp, *Artemia parthenogenetica,* the crustacean related to water fleas. Parthenogenetic species branched off from sexual species about 5.6 million years ago when the Mediterranean Sea was cut off from the Atlantic and dried up. Some Old World species consist only of females who *always* reproduce without male sperm. In these species, one female can colonize an empty pond. New World brine shrimp reproduce less exotically: sexually, with males.

Parthenogenesis—sexual reproduction without sperm—generally results in nearly identical clones. But some colonies of Mediterranean brine shrimp produce more than 60 genetic strains with a variety of lifestyles.

▶101 Ever felt tied up in knots? If so, think of the yellow-bellied sea snake. To molt and clean algae and barnacles off its skin, it ties itself in an overhand knot and pulls the knot from one end of its body to the other. How will this knot-tying sea snake become a harbinger of global warming?

The yellow-bellied sea snake, *Pelamis platurus,* perhaps the most numerous reptile on earth, avoids cold water. It lives only in warm regions of the Pacific and Indian Oceans from the Cape of Good Hope to the Panama Canal. If global warming raises ocean temperatures, the sea snake may move into the snake-free Atlantic Ocean. If it does, swimmers must be alert. The yellow-bellied sea snake is five times more venomous than the cobra.

▶102 Why do seagulls hang out at landfills?

A. Fast food.
B. Highly nutritious food.
C. Fun and frolic.

A and C. Some gull groups rely almost exclusively on dumpster left-overs, fried chicken, and barbecued ribs. Others prefer to catch more nutritious fish during the energy-intensive breeding season. But they all visit the dump to socialize. Sea gulls are highly social animals and rarely do anything alone.

▶103 Surprisingly, a young raven flies *away* after spying a moose carcass. A few days later, the bird returns with a gang of raucous adolescents to share the food. Why didn't the youngster hog the moose in the first place?

Adult ravens are fiercely territorial and viciously fight off young who try to share their food. Young ravens find food and fend off adult ravens by pooling intelligence about food finds and then forming ad hoc gangs of nine or more unrelated birds. More than 500 unrelated ravens have been seen converging from hundreds of square miles on one carcass. Ravens forage independently but sleep over at communal roosting places to share information about food finds. So was the first raven sharing the moose altruistically? Hardly. Young ravens cooperate for the common good for their own benefit.

▶104 Birds find tasty, camouflaged caterpillars by looking for plants with damaged leaves. How can the caterpillars hide from the birds?

By eating *very* neatly. Palatable caterpillars often mimic the leaves they eat *and* sweep their platters clean. They pare leaves evenly, leaving no rough edges. And they snip off half-eaten leaves, clearing all trace of leftovers. Wouldn't Miss Manners be pleased?

In contrast, many hairy, spiny, or poisonous caterpillars are messy eaters. They flaunt their presence by leaving tattered leaves and holes visible from afar.

Black-capped chickadees in aviaries quickly learn to check damaged leaves for tasty caterpillars. And bluejays can differentiate between leaves chewed by tidy tasty caterpillars and those mauled by unpalatable ones.

▶105 Waccamaw fatmuckets, fuzzy pigtoes, pocketbooks and snuffboxes, purple bankclimbers and Tennessee heelsplitters have irresistible names. But good names aren't everything, even in the advertising age. These creatures belong to the most endangered family of organisms in the United States. What is it?

Freshwater mussels. More species of the Unionidae mollusk family are candidates for protection under the Endangered Species Act than any other family. About 300 species—one-third of the world's—are native to the United States. About three-quarters are in serious trouble: approximately 20 considered extinct, 60 listed as threatened or endangered; and 70 proposed for protection. They particularly live in the Southeast; Alabama's streams and rivers, for example, are home to about 60 percent of known mussel species. The problem? Loss of habitat caused by stream damming, dredging, gravel mining; siltation from streamside logging and farming; and poisoning from sewage, toxic waste, and polluted runoff. Another problem: despite their names, they're not cute and cuddly.

► **106** On a happier note, what do a giant muntjac deer, a Vu Quang ox, an Indonesian tree kangaroo, and a Panay cloudrunner have in common?

All are new species discovered during the 1990s. The Vietnamese wilderness is home to the 100-pound muntjac, or barking deer, and to the wild oxen. The primitive tree kangaroo, as large as a midsized dog, has pandalike markings on its thick fur and hails from New Guinea. The cloudrunner, a 2-pound tree climber, is a nocturnal, squirrel-like rodent similar to a North American fox squirrel. All are threatened by habitat destruction.

► **107** On another happy note, how do woolly flying squirrels and Cebu flowerpeckers differ from Wollemi pines?

All three were thought extinct until their rediscovery during the 1990s. However, the last flying squirrel and flowerpecker bird were thought to have died early in the twentieth century, while the pine tree was believed to have disappeared 150 million years ago. The pine, known previously only by its fossils, is thus a living fossil.

The woolly flying squirrel is a 2-foot-tall squirrel, the largest living member of its family. It glides from boulder to boulder in the Pakistani Himalayas. The bird lives in a tiny forest patch on the island of Cebu in the Philippines. The 100-foot-tall pine was found with 41 others in Wollemi National Park, 125 miles west of Sydney, Australia. Their nearest relatives died in the Jurassic and Cretaceous eras. Habitat destruction, hunting, and collecting are threats.

Zoology

and Animal

Physiology

▶108 Sperm banks are nothing new. Many female birds and reptiles store sperm in their reproductive tracts for years before using it to fertilize their eggs. What's the record for banking sperm? And what female holds it?

A. Seven years.
B. Four years.
C. 117 days.
D. Two days.

A, the record, seven years, is held by the Javan wart snake (*Acrochordus javanicus*). Other snakes and turtles bank it for four or five years, while birds keep it for weeks. Domestic turkeys hold the avian record at 117 days. Human females are in the minor leagues at only five days, and domestic sheep and pigs are hardly in the running at all at two days.

Storing sperm allows females to fertilize their eggs at the best time with sperm from the most vigorous male, no matter when he was available.

▶109 Courting disaster, promiscuous female fruit flies die younger than their more demure sisters. What kills the frequent fliers?

Semen toxic to females. The male fruit fly's seminal fluid maximizes his mating success at the female's expense. Sound familiar? The fluid contains molecules that increase the female's egg production, destroy sperm from previous matings, and reduce her receptivity to future mating. The female, on the other hand, wants to mate only until the fertility of her eggs is certain. Each time she mates, she increases her exposure to the semen's toxins. Eventually, she pays the ultimate sacrifice: a shortened lifespan. All may be fair when love's a sperm war between the sexes. In this case though, it's the female *Drosophila melanogaster* that pays the price.

▶110 **Among zebra finches, the most adulterous males are those that have:**

(Pick one of the following.)

A. **The most and the fastest sperm.**
B. **The strongest and the fastest flight pattern.**
C. **The most and the least-functional plumage.**

A, the most and fastest sperm. DNA analysis of songbirds and their offspring has revealed that, in general, songbirds are a randy lot. While many males help their mates raise young, promiscuity is rampant and adulterous liaisons are disproportionately fertile. After a male zebra finch mates with his steady female, he sticks around to help her raise their young. During this period, the female is not sexually receptive, so the male spends his spare time looking elsewhere. In infrequent flings with other females, he produces seven times more sperm than he does with his regular mate and his sperm travels almost twice as fast. In fact, *extra*-pair copulations fertilize about half of all eggs, apparently because of this abundance of potent sperm. Hence a few adulterous liaisons produce more young than lots of day-in-and-day-out sex at home.

▶111 **To scan the ground for small rodents, kestrels soar high in the sky or perch on tall trees. Kestrels may travel more than 600 miles over unfamiliar territory searching for food. When rodent populations crash—as they do over broad areas every four years or so in the far north—locating the**

few surviving voles would seem well-nigh impossible. Yet kestrels manage. How?

Ultraviolet vision, the first discovered in a bird. The Eurasian kestrel (*Falco tinnunculus*) uses UV vision to quickly screen large areas for traces of voles (*Microtus agrestis*). These small rodents communicate with one another by blazing trails through grassy territories and marking them with urine and feces scent. Their scent chemicals absorb more ultraviolet light than the surrounding vegetation. When the scent marks fluoresce with ultraviolet light, a kestrel can easily spot vole habitat from far away.

Other raptors may have UV vision, too, because mouse urine also fluoresces in blue.

► **112** The Amazon mollie is the ultimate feminist fish. She—*all* Amazon mollies are female—naturally produces only daughters, and she does so by cloning herself without benefit of a male's genes. She needs male sperm though, as a chemical trigger to start her embryos growing. But where in an all-female species can she find a willing male? And what would be in it for him, other than a free fling with a philandering feminist?

She mates with the male of a closely related species, a sailfin mollie. Both sailfins and Amazon mollies are small, nondescript, gray fish that school in small streams throughout southeast Texas and northern Mexico. The name "Amazon" comes from the mythical tribe of women warriors, not from the South American river.

A male sailfin agrees to the charade because his affair turns on any female sailfin who sees him with an Amazon. To a jealous female, one mollie looks like any another.

The entire affair fascinates biologists. They used to think that females choose (male) mates for their manly attributes—for example, handsome tails or lithesome dances that speak to his overall health. Now they realize that some females lust for any male that another female wants.

► **113** Canary parents ignore polite children, and the meek do not inherit the nest. So when a parent arrives, an infant canary raises its head pugnaciously, opens its beak wide, and squawks for dinner. But the oldest and biggest chick in the nest isn't always the most aggressive; sometimes it's the youngest chick, the one just out of its egg. How can that be?

The mother canary laced her embryos with testosterone, the male hormone, even before her eggs were fertilized by a male. And she can give her first-laid egg the smallest dose and each succeeding egg a larger one. Thus the youngest may have 20 times more testosterone than the oldest. The hormone helps younger chicks compete with older siblings, but it can't outweigh the advantage of early hatching. Testosterone helps all the chicks develop body mass and feistiness and may improve their coordination by developing their spinal cords. Hormones play a vital role in development.

A newly hatched canary: The meek don't inherit the nest.

►114 Why do snakes have forked tongues?

A. "For picking the Dirt out of their Noses, which would be apt else to stuff them, since they are always grovelling on the Ground, or in Caverns of the Earth" (Hodierna, 1646).
B. To catch flies between the tines of their forks (Clayton, 1693).
C. For "a twofold pleasure from savors, their gustatory sensation being as it were doubled" (Aristotle, fourth century B.C.).
D. Because they are malicious and duplicitous.

C comes closest, although snakes actually have no taste buds on their tongues. Instead, forked tongues provide snakes and lizards with stereo smell for tracking prey and mates. First, the tongue collects molecules left behind by other animals. Then the snake's brain compares the strength of the chemical signal

The forked tongue of a reticulated python.

on one tongue tip with the signal strength of the chemical on the other side. In that way, the snake can locate and follow invisible scent trails left by passing animals. The paired ears of owls provide stereo sound in much the same way.

All snakes have forked tongues, and in some species of snakes and lizards the distance between their tongue tips is wider than their heads. Species with forked tongues tend to be highly proficient trail followers and foragers. Lizards with normal tongues are more apt to ambush prey.

▶**115** A barn owl's face is a model of abstract symmetry. Yet its ears are cockeyed or, to be more precise, cock*eared*. The left ear is higher than eye level and points down, while the right ear is lower than the eye but points up. Why?

The barn owl's left ear is more sensitive to sound from below, while its right specializes in sounds from above. As a result, each ear hears a sound at a slightly different time and intensity. The owl's brain analyzes these differences in separate circuits and then fuses the information to pinpoint the position of prey in the dark. Barn owls (*Tyto alba*) use timing differences to locate sounds in the horizontal plane and use intensity differences to locate sound vertically. How the bird calculates distance is not understood.

▶**116** Who are these complex creatures?

• The male is a househusband. He establishes territories, builds nests, and raises babies. He's not a homebody, though. Once he's courted a female and gained control of her eggs, he chases her away and cuckolds other househusbands.

• The female is literally a femme fatale. Her presence drives the male to an early death, and she ferociously cannibalizes the eggs of other females. As materialistic as a gold digger, she seems to favor only those males who elaborately decorate their nests.

Three-spine sticklebacks, *Gasterosteus aculeatus*. Sticklebacks are finger-length fish found worldwide in temperate climes. Thanks to their complicated sex lives, widespread availability, and adaptability to laboratory life, they are used to study how different behavioral and physical characteristics help animals adapt to their environments. For example, the female's preference for nests festooned with algae shows that, although she certainly doesn't intend to raise her young, she does at least want them camouflaged.

 What animal sports the densest fur?

A. Mink.
B. Sable.
C. Sea otter.
D. Seal.
E. Polar bear.
F. Chinchilla.

C, the sea otter. Blubberless mammals, they survive cold Pacific waters from north Japan to California by growing thick and double-layered fur. Its bottom undercoat has half a million hairs per square inch. Otters spend hours daily grooming their fur into prime insulating condition. For energy, they may eat 30 percent of their body weight daily. They are the largest of the mustelid group that includes freshwater otters, weasels, minks, skunks, and badgers. Fur hunters drove sea otters almost to extinction; early in the twentieth century, about 2,000 remained worldwide. With government protection, the sea otter population has rebounded to about 150,000, although in California they lag far behind, possibly because of pollution and fishing net accidents.

▶118 **What irresistibly cute and cuddly-looking animal routinely practices rough sex, kidnapping, theft, and even murder?**

The sea otter. A mating male holds onto a female by clawing her, biting her nose, or holding her underwater. In the process, some females are maimed or killed. Pups steal food from their mothers, and adult males may steal a third of their diet from females. Sometimes a male may even hold a pup ransom until its mother hands over her food. Oh, my. What PR agents they must have.

▶119 **With amatory gusto, midwestern prairie voles copulate hourly for 30 or 40 hours—continuing long after the female has become pregnant. What's the point of this mouselike creature's marathon of love?**

To turn on vasopressin, a hormone of fatherly love. During this first mating session, voles become that rarest of mammalian phenomena: a monogamous couple mated for life. Fully 97 percent of mammalian species are promiscuous and polygamous. Vasopressin, however, transforms normally timid virgin males into feisty fighters who defend their mates and territories and help rear their young. *Warning:* Vasopressin does not make female *or* male voles sexually faithful and true. DNA fingerprinting reveals that both males and females stray, so papa vole may be raising the offspring of another male.

Without an infusion of the brain hormone vasopressin, the male prairie vole (*Microtus ochrogaster*) might be as happily promiscuous and polygamous as its close cousins: meadow and mountain voles. On the other hand, giving meadow and mountain voles vasopressin does not make them monogamous; they lack the special brain receptors for vasopressin that prairie voles have.

Vasopressin's role in creating social bonds may help autistic children, who have difficulty forming social attachments, and schizophrenics, who become socially isolated.

▶120 **Here's a batty gender-bender for you: What *male* mammal makes breast milk?**

The fruit-eating bat *Dyacopterus spadiceus* from Malaysia, commonly called the Dayak fruit bat. It has an unusual pair of attributes: well-developed, lactating breasts *and* well-developed testes. It also has an 18-inch wingspan and a doglike face.

A few rare cases of milk-making male goats and sheep have been reported, but they were severely inbred domestic animals and may be a genetic mutation.

A lactating male Dayak fruit bat.

Why the male Dayak fruit bat lactates is not understood. It may sire *and* suckle young, or it may produce milk because it eats plants rich in estrogens. In any event, the male can't compete with a female Dayak in the milk department. He produces only a tenth as much milk. But perhaps he could get an A for effort.

▶**121** Female bats, common marmosets, elephants, bottle-nosed dolphins, and humans form a mammalian sisterhood of sorts. What sort is it?

All volunteer as midwives to help other females of their species give birth. For example, a female Rodrigues fruit bat "midwife" spent three hours assisting a first-time mother. The helper licked and groomed the mother's anal and genital area, helped her change position, groomed the baby as it emerged, and helped it crawl toward its mother's nipple. This fruit-eating bat, *Pteropus rodricensis*, lives in the wild only on the small island of Rodrigues, 500 kilometers east of Mauritius, and fewer than 350 remain there. The first bat midwife known to science lives in a captive colony in Gainesville, Florida.

The midwife Rodrigues fruit bat.

▶122 What's the fastest move in the animal world?

A. The pounce of a cheetah.
B. The blink of an eye.
C. The release of a jellyfish's stinger.
D. The jump of a click beetle.
E. The bite of a trap-jaw ant.
F. The jump of a flea.
G. The jump of a springtail.

The heads of two different trap-jaw ant species. The upper is Odontomachus bauri, *and the lower is the equally vicious* Daceton armigerum.

E, the bite of a trap-jaw ant, clocked at 0.33 millisecond. In comparison, the jellyfish was timed at 0.5 millisecond; click beetle at 0.6 millisecond; flea, 0.7 to 1.2 millisecond; springtails, 4 milliseconds; and the eyelid at a drowsy one-third of a second.

Trap-jaw ants, *Odontomachus bauri* from Arizona, northern Mexico, and Costa Rica, don't rely on muscle to snap their jaws shut. No muscle could work that fast. Instead, muscle cocks the ant's machete-blade jaws *open* and a tendon spring snaps them shut. Brushing against prey, touch-sensitive nerve endings on a hair trigger relay messages to the ant's brain via outsize, extrafast axon fibers. The noise of the jaws snapping shut can be heard a meter away.

▶123 What is the longest known incubation period for any bird?

Three months—by the *male* brown kiwi, *Apteryx australis*. Males do all the incubating in only a few bird species, but rearing a kiwi babe is a gargantuan task. Most birds need only one or two days to make an egg. But a 5-pound female kiwi needs a record-breaking month to make and lay one egg. No wonder: It weighs almost a pound, and it's four times more massive than an egg laid by the average bird its size. Moreover, 60 percent of the egg is fat-rich yolk.

After the female lays this superegg, her monogamous mate incubates it so she can recover. For three months, the male leaves the egg untended for only a few hours each night while he catches food. In fact, the male may devote 13 percent *more* energy incubating the egg than the female did laying it.

Why produce one or two supereggs instead of a clutch of small ones? Kiwi populations were dense and stable before Europeans brought dogs to New Zealand. Large hatchlings had a competitive edge over smaller nearby neighbors.

▶124 What bird has the only smell-sniffing *nose* in the avian world?

The kiwi. A flightless, nocturnal bird, it has nostrils near the tip of its bill to smell out and uproot earthworms and other soil organisms. For their highly developed sense of smell and hearing, their furlike feathers, and their low body temperature, they have been dubbed "honorary mammals."

Feral dogs have endangered kiwis, however. For example, an abandoned bitch killed roughly 500 kiwis in six weeks in 1987. That was half of one of the world's last sizable populations of kiwis.

▶125 What shocks oceanographers as they watch a torrid video of two deep-sea octopuses copulating?

Both octopuses are males of unknown and entirely *different* species.

The 16-minute video was taken from the deep-sea submersible *Alvin*, 2,512 meters below sea level on the East Pacific Rise. One small, white octopus draped itself atop a larger, brown octopus. Then the little octopus slipped its arm—the specialized, sexual arm characteristic of male octopuses—under its larger brown partner and inserted its armtip into the brown's body cavity. As the white octopus rapidly expanded and contracted its mantle, apparently boosting its respiration seven to nine times per minute, the biologists realized that the octopuses were "engaging in copulatory behavior."

The scarcity of octopuses in the ocean's depths may make males eager to mate with whatever happens by on the off chance that it's a female of the same species. Male octopuses are sexually mature for only a brief part of their short lives. An octopus who fails to take every opportunity offered may miss the chance of a lifetime to mate.

▶126 Seagulls, like most seabirds, drink salt water and eat sodium-rich food. So why aren't seagulls permanently pickled in brine like sauerkraut?

Because they have salt glands and drippy beaks. Marine birds such as gulls, petrels, albatrosses, terns, and sea ducks have a salt gland above each eye. When a bird's sodium content gets too strong for its kidneys to handle, the glands concentrate sodium chloride in solution and secrete it through ducts to the nasal cavity. From there, the solution drips—much like a small child's nose—off the tip of its beak. The petrel sneezes it out.

A seagull that ingests 10 percent of its body weight in salt water can excrete about 90 percent of the salt in three hours. The drip is roughly 5 percent salt—five times saltier than normal body fluids.

▶127 A baby loggerhead sea turtle hatches at night on a Florida beach, clambers down to the water, and finds its way—all by itself—hundreds of miles into the mid Atlantic Ocean. There it lives five to seven years before swimming back to nest on the very same beach where it was born. How does the loggerhead navigate?

A. Starlight and moonlight bouncing off the ocean.
B. The motion of the waves.
C. Magnetic fields.
D. Ocean currents.
E. Smell.

A through D, with E still to be proven. Emerging from their eggs, loggerhead turtles (*Caretta caretta* L.) head toward the bright light reflected off the ocean. To leave shore, they head straight into the waves. Catching a ride from the Gulf Stream, they ride currents north and east toward Portugal. At a fork in the current, they have a choice. They could head north to their deaths in cold water off England; instead, they use the earth's magnetic fields—how is not yet known—to swim south to the safety of the Sargasso Sea in the mid North Atlantic. There they spend five to seven years feeding on invertebrates.

A sea turtle's brain contains particles of lodestone, the magnetic mineral magnetite. Some mud-loving bacteria make internal compasses of magnetite to navigate; they use earth's magnetic field, the lines of magnetic force that flow around the planet from the magnetic south pole to the magnetic north.

▶128 What was the fiercest carnivorous predator to terrorize South America from 62 million to 2.5 million years ago?

A. A bird.
B. A mammal.
C. A marsupial.
D. A reptile.
E. A cat.

A, terror bird, *Phorusrhacos andalgalornis*. Its family, composed of at least 25 species, ranged in height from 3 feet to 9 feet. The tallest had a yardlong head, laid eggs the size of basketballs, *ran* 45 miles per hour, and used its small wings for balance, not flight. With a bladelike beak, it swallowed prey whole.

At the top of the grassland food chain 60 million years ago during the age of mammals, terror birds filled the niche occupied in North America by fleet-footed carnivorous cats. Terror birds had no natural predators. When South America became an island continent, terror birds reigned supreme over carnivorous marsupials and herbivorous mammals. The reign of the terror birds came to a close 2.5 million years ago when tectonic uplift and the buildup of polar ice caps made sea levels drop and the Panamanian bridge between North and South America reappear. As North American mammals swarmed south, they eradicated the terror birds.

▶129 Where did the earliest known ancestor of the human family come from?

From Afar, 4.4 million years ago. Fossils of the chimpanzee-like human ancestor were discovered in the Afar Depression of northern Ethiopia in 1992 and 1993. The new genus and species, *Ardipithecus ramidus,* is the ancestor of all later hominid species, including Lucy and human beings but excluding great apes, orangutans, and others.

People are one of the youngest major species on earth. Modern *Homo sapiens* emerged in Africa only 200,000 years, or 10,000 generations, ago. They differ from great apes in having 46, rather than 48, chromosomes. At some point a single block of ape genes was incorporated into an end of human chromosome 2.

▶130 Long praised as peaceful vegetarians, chimpanzees are now known to be avid meat hunters and eaters. Chimps eat meat during dry seasons for nourishment, but they hunt for other reasons, too. What are they?

Trading meat for friendship and sex, chimps hunt for social, political, and sexual reasons. The single best draw for a hunting party isn't meat: it's a female groupie in heat. Females given generous portions of meat after a hunt produce more surviving children. Most males in a chimp community are half brothers, and female chimps copulate with many of them.

Of the higher primates, only chimps and humans hunt and eat meat regularly. A chimp community can consume roughly 500 pounds of meat yearly, almost as much as some human hunting-and-gathering populations do today. In Gombe National Park in Tanzania, 75 percent of their food consists of baby red colobus monkeys. Male chimps make 90 percent of the kills, and they kill almost half the red colobus monkeys in their territories.

▶131 Besides people, what is the only animal that develops colon cancer on its own?

Cotton-top tamarins of Colombia. These 6-inch-high primates sport stylish punk hairdos, speak a grammatical language with 38 distinct sounds, and are cooperative and peaceful—among themselves, that is.

Childbearing is an immense burden on a cotton-top, so only one female reproduces in each community. Each year she produces twins that together weigh 25 percent of her body weight. (That's comparable to a 130-pound woman bearing and nursing 32 pounds of newborn twins.) Other cotton-tops suppress their own reproductive careers to care for her twins like their own. Cotton-tops who try to bear children without having had years of this child-care training wind up abusing, fatally dropping, or killing their young.

A 6-inch-high cotton-top tamarin of Colombia.

Cotton-tops are among the most endangered primates. In the 1960s and 1970s medical researchers imported 20,000 to 30,000 cotton-tops for colon cancer research. Fewer than 2,000 remain in the wild.

▶132 An animal that shares 98 percent of our genes lives in a female-dominated and egalitarian community. It has abandoned aggression for rampant, promiscuous sex—often in the missionary position. What is this everloving animal?

Bonobo, *Pan paniscus,* a great ape the size of a small chimpanzee, discovered in 1929. Genetically, bonobos are as closely related to us as a chimp or as a dog is to a fox. Humans split off from bonobos and chimps only 8 million years ago. About 10,000 bonobos live in humid forests south of the Zaire River. They have the graceful proportions of a Mannerist painting, with elongated legs, a small head, and narrow shoulders.

Bonobo sex is relaxed, casual, brief, and ubiquitous—more like handshakes and hugs between friends than a Hollywood love scene. Sex is a basic form of communication, used not to create tension but to diffuse it. Females are almost continuously sexually receptive and active. In fact, bonobos become sexually aroused at the drop of a hat and indulge in a variety of erotic behavior with almost every partner combination imaginable: from genito-genital rubbing to genital massage, intense tongue kissing, penis fencing, oral sex, and 13-second copulations face-to-face. Like many women today, bonobos have separated sex from reproduction: Female bonobos produce only one infant every five or six years.

▶ **133** What animal holds the world's growth record? Starting life as a larva one-tenth of an inch long, it grows 60 million times until it is 10 feet or more long and weighs 2.5 tons. If its larva were the size of a rowboat, an adult would be as big as 60 ocean liners.

An ocean sunfish, Mola mola, *from Everhard Home's* Lectures on Comparative Anatomy, *published 1814–1828.*

Giant ocean sunfish called molas. Huge relatives of puffer fish, they are the heaviest bony fish. Female molas can carry between 20 million and 50 million eggs, but a four-footer had 300 million—possibly another world record for this gentle giant. Molas lack true tails, so Germans call it a *Schwimmenderkopf,* or swimming head. Its Latin name, *Mola,* means millstone. It can bask on its side in the sun or swim 1,800 feet deep for moon jelly, squid, jellyfish, gelatinous plankton, small fish, crustaceans, and algae. Its tough, 6-inch-thick skin is reinforced with dense collagen fibers. An adult sunfish supports thousands of parasites—and parasites on parasites. Humans, killer whales, and sea lions prey on molas in tropical and temperate oceans.

▶ **134** Animal, vegetable, or fungi? is biology's basic question, and after that come only 35 or so animal phyla. In 1995 a new phylum was discovered: Cycliophora *Symbion pandora.* Its only member is a dust-buster creature the size of a pinhead. *Symbion pandora* has a suction cup at one end of its saclike body and a round upholstery brush affair at the other. Its reproductive system cycles elaborately between sexuality and asexuality. It rebuilds its feeding structures as they deteriorate but produces some larval types that never feed. What does this microscopic vacuum cleaner clean?

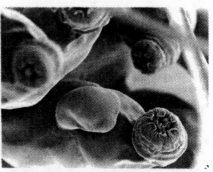

This microscopic vacuum cleaner, Symbion pandora, *occupies a new phylum all its own. In this feeding stage, its round mouth ring is closed and a budlike dwarf male is attached to its trunk.*

Lobster lips. *Symbion pandora* uses the tiny hairs around its mouth to tidy up after messy lobsters' meals. Its millimeter-long body was found on a slide stored for 30 years in a Danish museum. Lips of mature Norwegian lobsters—close kin to Maine lobsters—are caked with the critters. A new phylum was last introduced in 1983; it too involved a microscopic marine animal.

▶135 What fierce bird stalks feral rabbits through forests on foot? *Clue:* It is the only bird that breeds in both the Arctic and the Antarctic.

The skua *Catharacta maccormicki*, a gull-like predator and scavenger with a bad reputation and a wingspan of 5 feet. One of the most aggressive birds in the world, skuas disgust many observers. For example, brown skuas are the primary predator and scavenger on Enderby Island between New Zealand and Antarctica. Because the island has no carnivorous mammals, skuas tidy up the local sea lion rookery by eating placentas and feces. Skuas steal penguin eggs and kill penguin chicks. Perhaps skuas have been maligned, though. They are the only predator keeping the island's large rabbit population in check. Sealers and explorers brought rabbits to Enderby to provision passing ships. And recent studies show that skuas have remarkably little impact on penguin numbers or breeding success. In fact, penguins destroy more skua eggs than vice versa.

▶136 What vertebrate has the most energy-hungry brain of all? *Hint:* Most vertebrates devote 2 to 8 percent of the oxygen they breathe to their brains.

A. A human.
B. An elephant nose fish.
C. A meadow vole.

B, an elephant nose fish uses 60 percent. Humans, long thought to be the only exception to the 2 to 8 percent rule, use only 20 percent. The elephant nose fish, *Gnathonemus petersii,* is nocturnal and lives in rivers and lakes in west and central Africa. Its modified muscle cells generate an electric field around its body, and unique electrical receptors detect changes in the field. Objects cast electrical "shadows" on its body surface, enabling it to navigate in the dark. Most fish have tiny brains, but a 10-inch-long elephant nose fish has as big a brain as a mammal its size. While the brains of most fish account for less than one percent of body weight and human brains are roughly 2.3 percent of body weight, the brain of an elephant nose fish accounts for 3.1 percent of its body weight. The animal that devotes the largest share of its body to brain tissue is the squirrel monkey; about 5 percent of its body mass is brain.

▶**137** What ancestor of both humans and great apes is named after a pipe-smoking, bicycling vaudeville performer?

Proconsul, the last common ancestor of great apes and humans. Proconsul was a slow-moving, cautious, arboreal ape that lived in Kenya about 18 million years ago. Roughly the size of a female baboon, Proconsul had not yet adapted to leaping, arm-swinging, knuckle-walking, or living on the ground. Parts of more than nine Proconsul skeletons have been found over the past 30 years. When the first Proconsul fossils were discovered in 1927, a popular vaudeville troop in London featured a bike-riding chimpanzee named Consul. As a joke, the newly discovered ape species was named for Consul.

▶**138** It's a wise person who knows which side his bread is buttered on, the old adage says. So which animal remembers best where it got its last meal—and how to find it again? Cats, chickens, dogs, goats, opossums, pigs, rabbits, rats, or turtles?

Opossums (*Didelphis virginiana*). Although marsupials may have originated in North America 100 million years ago, the only one around today is the opossum. Its prowess at locating food has helped it to survive in modern times and even to expand its range. Nearly 80 species of marsupials—30 percent of the world's total—live in Central and South America.

A DINOSAUR QUIZ

Dinosaur science entered a golden age about 1975. As a result of new discoveries and technology, almost everything previously known about dinosaurs has been reexamined. One question, in particular, has been challenged: Were dinosaurs cold-blooded reptiles? Check how up-to-date you are on dinosaur lore by answering the following questions.

▶139 Dinosaurs were "lay-'em-and-leave-'em" parents. (True or False)

False. Fossil evidence suggests that some dinosaurs may have excelled as protective parents at least during the last 10 million years of their reign on earth. The fossil of a dinosaur sitting on her eggs was found in Mongolia, and several nests with eggs and babies in Montana indicate maternal care, too.

▶140 Dinosaurs were gray. (True or False)

False. Rhinos, hippos, and elephants are gray because they are color-blind. But all living reptiles and birds—descendants of the dinosaurs—see in living color. Dinosaurs had large eye sockets and optic lobes, another indication that they were probably visually and color oriented. And like birds, dinosaurs may have been adorned with bright colors and skin frills for sexual display and easy herd recognition from afar.

► **141** Dinosaurs evolved slowly. (True or False)

False. Turtle and crocodile species have kept the same body shape for at least 100 million years, but dinosaur species evolved and disappeared every 2 million to 4 million years, and entire families of species could disappear within 20 million years.

► **142** Being cold-blooded reptiles, no dinosaur could survive a dark Arctic winter. (True or False)

False. First, the Arctic was dark but not cold during the dinosaur age. The whole world was like a global greenhouse, and fossils of elephant-sized and larger dinosaurs have been discovered even on Alaska's North Slope. Second, while some dinosaurs may have been cold-blooded, large dinosaurs may well have been warm-blooded. Most large dinosaurs were elephant-sized or bigger, and all multiton land animals today are warm-blooded, even at the equator. If their body temperatures dropped even 5 degrees Fahrenheit, they would lose 30 percent of their muscle and nerve performance. Many dinosaurs grew extremely rapidly. During their last 10 million years on earth, some grew from chicken-sized infants to large adults in about two years; such rapid growth indicates warm-bloodedness. Their bone structure, adapted for fast movement, requires warm-bloodedness. And dinosaur bones resemble the bones of living mammals more than those of reptiles.

► **143** Smart mammals exterminated big, stupid dinosaurs. (True or False)

False. Some dinosaurs had big brains, and small dinosaurs evolved brains faster than mammals did. During the 160 million years of the age of dinosaurs, dinosaurs outclassed mammals so consistently that mammals as a group could not evolve a single species bigger than a woodchuck.

▶144 **Dinosaurs died out all at once worldwide. (True or False)**

False. Five or more "extinction events" occurred over 160 million years of dinosaur history. Each time, those that died out were large, active land animals. A dozen mass extinction events have occurred throughout history, and their typical targets were advanced species that had evolved to survive in special situations; "generalist" or "opportunistic" species, which could adapt to a variety of situations, survived. After the last extinction event, which did destroy the dinosaurs, the survivors were small mammals, frogs, salamanders, and lizards on land and large turtles and crocodilians in fresh water. Large, fast-evolving species are vulnerable to the sudden extinction of all or most of their species.

▶145 **Dinosaurs are extinct. (True or False)**

False. Today, 8,300 to 8,700 species of them are alive: Birds. According to overwhelming skeletal evidence, birds descended directly from a small, meat-eating dinosaur. To repeat: most dinosaurs were probably not cold-blooded.

Plant Science

▶146   Which came first: the insect or the flower?

The insect. Insects have evolved 34 different kinds of mouthparts, depending on whether they bite, cut, tear, sieve, suck, pierce, chew, grasp, lick, or sip their food. And 85 percent of those mouthparts developed roughly 100 million years *before* the spread of flowering plants. Fossils from 1,263 insect families—including Russian and Chinese fossils unknown to Western scientists until recently—suggest that plants grew flowers to lure insects into pollinating them. Scientists had assumed the reverse: that insects evolved in response to the emergence of flowers 125 million years ago.

▶147   What and where is the oldest *living* fungus known to science?

In the hay insulating the boots worn by Ötzi, the Bronze Age alpine hunter frozen 5,300 years ago. Ötzi was discovered in melting snow in the high Alps in 1991. Dormant spores of microscopic fungi *Absidia corymbifera* and *Chaetomium globosum* were found in his boots. Placed in a laboratory dish after 60,000 generations of dormancy, they sprang to life. Scientists believe the spores were in Ötzi's boots before he died, because neither fungus grows at high altitudes.

▶148 In a model of cooperation, each yucca species comes with its own personal moth species. The yucca gives its moth exclusive rights to pollinate its flowers. In return, the yucca sacrifices some of its seeds to feed the moth's caterpillars. But what if the moth gets greedy and infests too many yucca flowers? How can the yucca protect itself?

Become an abortionist. Abort the grub-infested flowers early in their development. While greedy moths lose all their offspring, the yucca produces hundreds of other flowers and barely notices its loss. Some model of harmony, eh?

A female yucca moth pollinating a yucca flower.

▶149 The Roman Empire's population declined gradually from 32.8 million to 27.5 million during the 500 years after Christ's birth. This mysterious drop in birth rates occurred without major wars or epidemics and despite a large immigration of German tribes. Which of the following made the empire's population fall?

A. Lead-plated eating vessels.
B. Infanticide.
C. Contraception and early-term abortions.
D. Food shortages.
E. Coitus interruptus.

B and C: infanticide to limit the number of women, and contraception and abortions induced by botanical drugs. Birth-control herbs were used effectively and were socially acceptable throughout the ancient world. For example, Greek and Roman women drank silphion plant sap from the giant fennel family to prevent conception and induce abortions. Silphion was so popular that it cost its weight in silver and, by the fourth century A.D., had been harvested to extinction. In studies, almost all rats fed modern extracts of its close relatives within three days of mating failed to become pregnant.

As for the other choices, lead poisoning had only a small effect and food supplies were adequate. Documents freely describe many sexual practices but there is no evidence that coitus interruptus was widespread; it depends on the male's cooperation.

Greeks and Romans understood the difference between contraceptives and abortion-inducing drugs. Furthermore, some believe that the ancient Hippocratic oath literally forbade only abortive suppositories, not abortions in general.

▶**150** Zero population growth is supposed to be a modern-day concept. But historically, the population of Tibet has often failed to grow for periods as long as 200 years. How?

Peas. Traditionally, Tibetans survived periods of hardship by eating primarily peas and barley. Laboratory mice fed a diet that was 20 percent peas halved the size of their litters. When 30 percent of their diet was peas, the mice failed to reproduce at all. Peas are only one of approximately 450 plant species known to contain natural substances that reduce fertility by preventing ovulation, fertilization, or implantation.

▶**151** Like damsels in distress, plants under insect attack can SOS for rescue. The white knights, in this case, are parasitic wasps that attack the insects. But how do wasps find one beleaguered plant in a field? And how does this wasp-plant conspiracy control caterpillars?

Parasitic wasps *learn*. Hours after a caterpillar bites into one of its leaves, the plant releases volatile hydrocarbons that pregnant parasitic wasps home in on. The wasp lays eggs in the caterpillar, and the eggs develop into larvae that gobble up the caterpillar.

The wasp learns to associate the plant's volatile hydrocarbons with the odor of caterpillar feces. Parasitic wasps can be trained to respond to other scents, colors, and shapes, too. No doubt, someday patrols of parasitic wasps will be trained to attack on command to protect innocent crops from the onslaught of ferocious plant-eating caterpillars.

▶152 How can an enterprising plant thrive even in a stunted, nutrient starved forest on a sandstone hill?

By forming an ant-plant cooperative in Malaysian Borneo. The plant—epiphyte *Dischidia major*—grows leaf sacs that provide cradle-to-grave shelter for colonies of *Philidris* ants. In return, the plant's houseguests have only to breathe, eat, defecate, and die. The plant gets 39 percent of its carbon needs from ant breath and 29 percent of its nitrogen from their garbage dump (ant feces, dead ants, and scavenged insect parts).

Dischidia major, *a B&B for ants.*

▶153 How is sorghum like the golden goose?

Within 24 hours of a grasshopper's chomping on a sorghum plant, the plant starts growing new leaves. The grasshopper secretes a chemical that stimulates the sorghum to serve its never-ending banquet. The saliva of grazing bison and mice encourages their dinner plants to regrow, too. The stimulating chemical is apparently similar to the epidermal growth factor (EGF) found in vertebrate saliva.

▶154 Anemones, tulips, buttercups, and poppies all bloom a dazzling scarlet in the Judean desert near the Jordan River. And they all have bowl-shaped flowers with black centers. Why?

They depend on the same beetle for pollination, and it sees red. Israel's "poppy guild" includes 15 species of large, red, cup-shaped flowers from three plant families. All rely on scarab beetles, *Amphiocoma*, that don't distinguish scents or shapes well but that do recognize red-carpet treatment when they see it. As far as male beetles are concerned, red means bed and breakfast, red-light-district style. Racing from blossom to blossom, they spend about three minutes in each; and the instant they find a female inside, they slam-bang mate. Females, on the other hand, linger an average of 16 minutes in each flower, methodically eating pollen for the protein their eggs need.

Flower members of the poppy guild get a good deal, too. While the average bee makes off with only 100 pollen grains per visit, a scarab beetle can carry approximately 2,000 pollen grains.

▶**155** Horseshoe crabs, crocodiles, and coelacanths are celebrated living fossils. They evolved so slowly that they resemble true fossils from 100 million years ago. But one type of living fossil is much more common and much older—between two and ten times older. What is the oldest known living fossil?

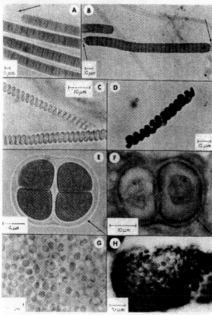

The cynobacteria in the left column are living; their fossilized counterparts appear opposite each on the right.

Blue-green bacteria, aka cyanobacteria or pond scum. They are virtually identical to billion-year-old fossils. Approximately 90 species of fossilized blue-green bacteria have contemporary look-alikes. Today's cyanobacteria thrive in exotic locales: hot springs, snowfields, and deserts; areas with little or no oxygen; and in water that is harshly acidic, salty, basic, or pure. A liter of seawater may contain 10 million cells of them. Cyanobacteria produce about 20 percent of the nutrients at the bottom of the marine food chain. With this much success, why bother evolving?

►**156** This plant grows only two leaves—but compensates by letting them live up to 1,000 or 2,000 years. The plant also enjoys a highly irregular sex life. In all other known plants, sperm travels to the egg for fertilization. In this plant, the egg migrates, moving into a pollen tube to unite with the sperm. What is this miraculous plant?

Welwitschia mirabilis. It lives in the oldest desert on earth, the Namib Desert, which stretches 1,300 miles along the Atlantic coast of southwest Africa. Because the plant relies on morning fogs for moisture, it grows only 100 to 200 millimeters yearly. Over their lifetime, the leathery green leaves can extend

Collecting insects under a Welwitschia *in Angola.*

10 to 20 feet and offer shelter to desert birds, animals, and insects. Its deadwood crown can get as big as a washtub. *Welwitschia* plants are either male or female and are pollinated by wind.

►**157** What plant produces the world's smallest flowers and fruits? A bouquet of blooms fits on the head of a pin, while the fruits are smaller than a grain of salt. *Hint:* Unlike other flowering plants and ferns, this plant has neither stems nor leaves. Some of its species are even rootless.

Duckweed, with 34 species (in the *Lemnaceae* family) throughout temperate and tropical regions. Some *Wolffia* species bud a new daughter plant every 30 hours, at a rate that could fill planet earth in four months. Given a few summer days, even ordinary duckweed can cover a pond, purifying its water and controlling mosquitoes. Dried duckweed is about 40 percent protein, much like soy beans. Some ducks get half their summer diet from duckweed. Ancient Mayans ate it; so do Thai peasants. Theoretically, 10 acres of duckweed could supply 60 percent of the nutritional needs of 100 dairy cows, and one acre could produce 3,000 to 5,000 gallons of ethanol yearly. Because it absorbs genes easily and produces genetically altered clones, duckweed could make tiny genetic factories producing pharmaceuticals and fertilizers. Look for it soon in your local supermarket.

▶**158** What is the world's largest single-celled organism?

A 2- to 3-foot-long *Caulerpa* plant, whose 73 species grow in warm and shallow seas from Italy to the Florida Keys. *Caulerpa*s are the most differentiated, single-celled organisms in the world. A *Caulerpa* looks like an ordinary seaweed, but its yardlong horizontal stems sprout leaves up and roots down. Yet each plant is a single cell; no cell wall or membrane separates leaves from roots or roots from horizontal stems. Other plants as large and complex as *Caulerpa* consist of hundreds of thousands of microscopic cells, with walls to hold them erect. A *Caulerpa* is held upright by buoyant seawater and by the single cell wall enclosing the giant cell.

*Caulerpa*s can change the placement of their leaves and roots faster than any other plant known, too. Turn a *Caulerpa* upside down, and new roots grow down within a day. When new leaves appear, they grow up.

▶**159** What is the world's largest cactus?

A. Cardon
B. Organ pipe.
C. Saguaro.

A, the cardon of Baja California and Mexico's western Sonoran Desert. Cardons may grow 60 feet tall and live more than 100 years. Like saguaros, they have a single trunk and multiple branches. Cardons, however, are more massive. They grow slowly and do not become sexually mature until they are 50 years old or more.

▶**160** Depending where you look, some cardon cacti are male and produce only pollen while others are female and produce only ovules. Still other cardons are traditionalists: like most flowering plants, they are her-

maphrodites, producing both pollen and ovules. What accounts for the cardon's complicated sex life?

Its proximity to female bats. Cardon cacti and lesser long-nosed bats
evolved together. The bats—which have a 13-inch wingspan but
weigh less than an ounce—mate in central Mexico. Then the pregnant females fly north alone, following the cardons' blooming cal-
endar. Up to 100,000 females may congregate in a cave or
abandoned mine, each female producing only one offspring yearly.
At night the mothers make four-hour foraging expeditions, searching for the cardon's white, night-blooming blossoms whether 15 or
50 miles away. The bats, totally dependent on plants for food, use
the amino acids in cardon pollen to make protein. In return they
pollinate the cardon. Male, pollen-producing cardons are so dependent on bats for pollination that the trees grow only within 50 miles
of a lesser long-nosed bat nursery.

▶**161** What plant was called "holy herb," "cure-all herb," and "holy Indian
cure" during the sixteenth century?

Tobacco. Europeans learned about smoking tobacco from American
Indians, who believed it had healing properties. Tobacco was introduced into Europe in 1556 as a medically beneficial substance, and
Portuguese and Spanish sailors spread it worldwide. Not until the
1950s and 1960s were its harmful effects documented and publi-
cized.

▶**162** What popular American product can be flavored with the following
additives? Extracts of anise, cinnamon, molasses, dandelion roots,
and walnut hulls; juice from apples, raisins, figs, and plums; black currant
buds; peppery capsicum oleoresin; clover tops; nutmeg; vanilla; vinegar;
smoke flavor; tea leaves; orange blossom water; and oils of basil, bay leaves,
caraway, carrots, dill seeds, ginger, lavender, lemon, lime, pepper, Scotch
pine, and oak chips; butter, chocolate, caffeine, coffee, cognac oil, cocoa,
honey, rum, water, sherry, and yeast.

Cigarettes. Reacting to congressional criticism in 1994, six tobacco
companies published a list of 599 flavorings that can season a

smoke. That doesn't include "processing aids" and 29 unidentified ingredients. Amounts used were not revealed. A previous, highly confidential list of cigarette additives included heavy metals, pesticides, and insecticides and at least 13 ingredients not permitted in foods, according to one congressman. With all those goodies, why doesn't everyone want to smoke?

▶**163** Tropical forests produce the largest number of species and rank first worldwide in species diversity contests. But what takes second place? What type of climate produces the next-largest number of species?

Places with big temperature changes between winter and summer. An enormous study compared plant diversity in 94 locations, from Australia's outback to Russia's Arctic. Conclusion? In midlatitudes, more plant species are found in areas with big seasonal temperature fluctuations than in areas with more even climates. The most botanically diverse regions of the earth seem to be well-lit and well-watered, assuring ample photosynthesis.

Many ecologists had hypothesized that equable climates with relatively even seasons would have the most biodiversity. But the plant survey suggests that the tropics are biologically rich despite their even climates, not because of them.

▶**164** Potatoes, tomatoes, chile peppers, peanuts, chocolate, and sweet potatoes originated in Central and South America. Where did pineapples, cashews, and avocados come from?

From South America. In fact, eight of the world's top 26 crops by tonnage first grew in the Americas. A third of the value of United States crops comes from plants of American origin. More than half the world's calories come from just two Latin American plants: potatoes and corn. In comparison, the United States contributed only cranberries, blueberries, pecans, black walnuts, sunflower seeds, and Jerusalem artichokes—hardly important staples in the world's diet.

▶**165** Today, North American farmers raise approximately 250 varieties of potato. How many varieties did pre-Columbian Incas grow? *Hint:* Terraces up steep mountainsides can include hundreds of microhabitats reaching as high as 15,000 feet above sea level.

A. 30.
B. 300.
C. 3,000.
D. 30,000.

C, 3,000. The Incas farmed more than 70 crops in the Andes and fed 15 million people from Colombia to Chile.

▶166 Napoleon said that an army marches on its stomach. What did the Incas' soldiers march on during their long, cold winters?

Freeze-dried potatoes. The potatoes set outside froze overnight. When they thawed the next day, workers stomped on them to remove excess water. Several days of freezing, thawing, and stomping produced an easy-to-carry food for marching Incan soldiers.

▶167 Most of the foods raised by the Romans 2,000 years ago originated outside Italy in parts of Eurasia. Match the following crops with their native habitat:

1. Oats, poppies.
2. Cucumbers, sesame, citrus fruits.
3. Quince.
4. Chicken, rice, apricots, peaches, millet.
5. Wheat, barley, lentils, peas, and domesticated sheep, goats, pigs, and cattle.
6. Millet, cumin.

A. Caucasus.
B. Central Asia.
C. Tigris and Euphrates Valleys.
D. Italy.
E. China.
F. India.

1–D, 2–F, 3–A, 4–E, 5–C, 6–B. Domesticated food crops from the Tigris and Euphrates Valleys spread rapidly east and west across Eurasia. Similar latitudes share similar seasonal variations in day length and often have similar temperatures, rainfall, and diseases. Climate varies more north and south than east and west. By the birth of Christ, cereals from the Fertile Crescent spanned 10,000 miles from Ireland to Japan across the longest land distance on earth.

▶168 The idea of farming began in Babylonia and Mesopotamia. Between 8000 B.C. and 4000 B.C., farming spread rapidly, first to Greece and Cyprus, then to Egypt and India, central Europe, and finally Britain. Did farming spread to other regions, whose people then domesticated their own local plants? Or were the actual seeds domesticated in Southwest Asia transported across Eurasia?

Surprisingly, the seeds themselves. Most early crops—barley, wheat, peas, and legumes—started with imported seed and shared one common genetic ancestor. Beginning farmers may have found it easier to start out with a neighbor's tried-and-true seed than to domesticate local relatives of the same plant.

▶169 Tall fescue is loved but lethal. It's the most popular pasture and turf grass in the central and eastern United States. It covers 35 million acres of parks, playgrounds, athletic fields, lawns, rights of way, pastureland, and so on. Tall fescue has a multitude of virtues: vigor, pest resistance, and tolerance of humidity, drought, and lousy soil. Yet tall fescue can poison livestock. What one thing makes tall fescue so valuable—and yet so dangerous?

Why do these cows have rough hair coats? Why are they suffering from the heat? The culprit is tall fescue.

An internal fungus, *Acremonium coenophiolum*. It makes the grass germinate better, resist drought, and produce more seeds. Its toxicity makes insects shun the grass. Unfortunately, this same toxicity—apparently caused by ergot alkaloids—poisons grazing animals. As a result, the animals do not gain enough weight, cannot tolerate heat, get gangrenous feet in cold weather, produce scant milk, and may have reproductive problems. Fortunately, the animals fully degrade the alkaloid so their meat and milk do not harm people.

The fungus evolved with its host plant to their mutual benefit. But it infects most of the tall fescue grass in the United States because most tall fescue seed derives from one infected Kentucky pasture. Today, the fungus costs livestock owners in the United States between $500 million and $1 billion yearly. The problem will be hard to solve because tall fescue (*Festuca arundinacea* Schreber) is the only cool-season perennial grass known to flourish in the south-central states.

Many other grasses also harbor internal fungi, but little is known about their impact.

▶170 Where does all the carbon go? Fossil fuel combustion and deforestation release into the atmosphere approximately 6 billion to 9 billion tons of the greenhouse gas villain yearly. Of that, 3 billion tons remain in the atmosphere, 2 billion tons are absorbed by oceans, and large amounts are absorbed by tropical forests and tundra. But that still leaves one billion to 2 billion tons of carbon missing and unaccounted for. Can you get to the root of the mystery?

It may be underground in tropical grasslands in South America. Pastures there may store between 100 million and 507 million tons of carbon yearly. Since 1980, South American ranchers have planted almost 90 million acres with deep-rooted African grasses that store 13 percent more carbon than neighboring savannas. Fields planted with African grass *and* legumes store 36 percent more carbon than savannas. Their roots may be absorbing some of the carbon released by tropical deforestation.

▶171 Nearly 300 horse-drawn threshing machines exploded and burned while harvesting crops in a single county of Washington State early in the twentieth century. What—or who—blew them up?

A. The Women's Christian Temperance Union.
B. The Wobblies.
C. Anarchists.
D. Smut.

Smut growing out of an ear of corn.

D, smut, specifically corn smut, *Ustilago maydis,* a black fungus that grows on corn. Smut forms bluish-gray boils nearly an inch long on corn kernels. When the boils break, black "soot" billows out in "smut showers" over large areas. Smut dust is more explosive than coal dust, and static electricity or sparks can ignite it. Smut damaged Greek and Roman grain crops, and in 1918 smut destroyed as much corn in the United States as the states of Indiana and Maryland grew. Approximately 1,200 species of smut infect 4,000 species of cereal crops, beans, dahlias, gladioli, and violets.

Smut was used as a hair and skin darkener in China and Japan and is a food delicacy in Mexico. Evidence of its toxicity in humans is inconclusive.

▶172 Kudzu, the fast-growing vine that ate the South, can grow more than a foot a day. Who or what is "Son of Kudzu"?

An unrelated Asian immigrant vine known variously as mile-a-minute weed, minute weed, tear-thumb (from the barbs on its stems), or *Polygonum perfoliatum.* Mile-a-minute grows half a foot a day, kills vegetation in its path by forming tangled thickets 20 to 25 feet high, and grows from seed over disturbed soil, parks, gardens, stream banks, and moist lowlands. In its haste, it scrambles over wildflowers, grasses, shrubs, and small trees. Now an annual south to Virginia, it could become a disastrous perennial in Florida. It is a major pest from Korea to India. Even in its East Asian homeland, it has no known biological enemy.

More than 4,500 imported species of plants and animals have found happiness in the United States. Of these, 79 have caused more than $97 billion in damage. Kudzu alone accounts for $50 million in lost farm and timber production each year. And "Son of Kudzu" intends to outdo that.

▶173 The rarest tree species in North America is hardly a household word. It doesn't have the panda's lovability, the eagle's nobility, or the dogwood's landscapability. Its very name is offensive. Nevertheless, its numbers have dropped dramatically since the 1950s, and only 1,500 trees remain. What's it called?

The stinking yew, because its needles reek. Better known as the Florida torreya, it grows along a short stretch of the Apalachicola River between Florida and Georgia. It is a close relative of the Pacific yew, producer of the anticancer drug taxol. However, a deadly fungus may kill the stinker off before scientists can find out if it too makes useful compounds.

▶174 Why should woodwind players refrigerate their reeds with their food and drink?

Refrigeration slows the bacterial growth and chemical decomposition that destroy reeds. Clarinet, saxophone, oboe, and bassoon players use reeds that are botanical crystals grown by the giant reed plant *Arundo donax*. The cellulose core of a reed vibrates because it is ordered—like a crystal—in a regular pattern. Gummy, moisture-retaining chemicals keep the crystal flexible.

Finding and maintaining a good reed is an expensive, time-consuming business for musicians, however. Their saliva is alkaline and quickly destroys the chemicals that keep the crystal flexible. When streptococcal bacteria encrust the reed's interior, they choke off vibrations. Finally, in the stress of being played, the cellulose in the reeds develops microfine cracks. A woodwind musician may spend more than $150 for 100 reeds, but find only two that last a few weeks.

▶175 A paramilitary campaign mobilized American schoolchildren in the Midwest to eradicate a popular garden shrub threatening the nation's wheat crops during World War I. Schools awarded medals to the children who located or destroyed the most specimens. Name the enemy bush.

Common barberry, an attractive shrub with sharp spines and scarlet berries. The plant is an intermediate host of stem rust disease, which devastates wheat. The children's campaign eradicated most of the barberry plants from the wheat-growing areas of North America and brought the disease under control. Many barberry varieties are still banned from the area.

▶176 Among the hardy mountain climbers scaling Alpine peaks these days are _____. They are forging upward at a heady pace of 1 to 4 meters per decade. Fill in the blank, please. And explain what they'll do when they reach the top.

Plants. Probably because of global warming, Alpine plants are climbing peaks to escape warmer climes below. A survey of 26 Alpine summits revealed a wholesale migration of plants skyward. The problem? They can't climb fast enough. Over the twentieth century, the mean annual temperature of the Alps warmed 1.0 to 1.5 degrees Fahrenheit. To escape the heat, the plants should be migrating 10 meters a decade, yet some can manage only one. Anyone know how to hurry up a plant? Of course, even if they did climb fast enough, where would they go once they reached the top of the mountain?

▶177 Ninety percent of this Latin American ecosystem has been destroyed in the past 30 years. Name the ecosystem, please.

Mountain cloud forests in the northern Andes. In comparison, only 20 percent of Amazonian rain forests have been destroyed. Mountainous forests grow between the Amazonian lowlands and alpine grasslands. While rain forests have more tree species, mountain forests are richer in herbs, bushes, epiphytes (plants growing on plants), and mosses. The Andes are one of only 12 places on earth where major food crops originated, and its plant diversity is caused by wide differences in humidity and altitude.

Thanks to urban migration, the northern Andes are now home to 70 million people, the world's largest high-altitude population. Mountain cloud forests have been denuded in Ecuador by gold mining; in Costa Rica by cattle ranching; in Bolivia by coca for cocaine; and in Colombia by coca and opium poppies.

▶178 A 2.5-acre plot of land in Peru contains more tree species than any other area its size on earth. How many types of trees were counted there? *Clue:* Together, the United States and Canada host 700 native species of trees.

A. 25.
B. 50.
C. 100.
D. 200.
E. 300.

E, 300. A biologist counted them in a rain forest near Iquitos, Peru. Second place may go to a 25-acre site in Borneo with more than 1,000 species.

▶179 Flesh-eating plants, such as the Venus's-flytrap, pitcher plants, and the aquatic bladderwort, are rare and bizarre evolutionary oddities. In fact, they defy the normal order of the food chain. Right?

Wrong. The fact that approximately 300 species of insect-eating plants have evolved demonstrates that meat consumption is an excellent survival strategy for plants as well as people.

▶180 Cuddly rabbits, twittering sparrows, and feral horses are charming creatures that can become downright pestiferous when introduced into new locales around the world. Interloping insects are rarely regarded as fondly, yet when introduced into new habitats, they have sometimes been beneficial. When and where was the first insect imported to kill weeds?

In India in the 1880s. The American prickly pear, *Opuntia vulgaris*, became a weed in India during the nineteenth century. It was cut down to size when its natural insect foe, *Dactylopius ceylonicus*, was called to the rescue. Worldwide, insects have been introduced into new habitats to fight more than 30 weed species.

▶181 What do "odor-eating" shoe liners, nuclear power plants, range hoods, smoke-eating ashtrays, gas masks, and Japanese cigarette filters have in common?

Activated charcoal made of coconut shells. It has fine pores less than 2-billionths of a meter in diameter. Pores of activated carbon made from wood, peat, or coal are tens or hundreds of times larger. Thus wood and peat carbons are used for liquids, but coconut shells are for gases.

The coconut's shell, incidentally, is not to be confused with its husk. The shell is located between the husk and the flesh.

▶182 The bark of the rare Pacific yew tree contains taxol, an antitumor treatment for ovarian, breast, and lung cancers. What other part of the yew also makes taxol?

A fungus discovered in 1991 growing under the bark of a yew tree in an old-growth cedar forest in northern Montana. Both the host tree, *Taxus brevifolia,* and its fungus, *Taxomyces andreanae,* produce the hormone taxol. Rice plants and a fungus that grows on rice also make an identical hormone; theirs regulates the growth and development of the rice.

Pacific yews—and their fungus—grow in the moist underbrush of old northwestern forests. Before chemists learned how to manufacture taxol synthetically, three trees had to be killed and stripped of their bark to treat each patient.

Earth Sciences

▶183 Earth's climate has been unusually stable for 10,000 years. In fact, our climate has been more equable and consistent than during any other 10,000-year period in the previous 100,000. Thanks to our balmy temperatures, civilizations could develop all over the planet. What stabilizes our era's climate?

Ocean circulation and the absence of big ice sheets on Canada and Scandinavia are probably the main reasons. The oceans transfer heat and salt around the globe in giant conveyor systems. The Atlantic Ocean moves warm water north from Antarctica to Greenland. There it cools, sinks, and returns to Antarctica. Warmer and less dense than Antarctica's frigid surface water, the Atlantic conveyor rises again and starts its cycle anew. Its northward current carries 15 times more water than all the world's rivers and is about eight degrees warmer than the southbound current. This enormous transfer of heat warms northern Europe.

If the oceans' conveyor belt stopped, winter temperatures bordering the North Atlantic could drop 5 to 15 degrees or more. The conveyor might restart only hundreds or thousands of years later.

►184 How fast can climate change?

A. In three to five years.
B. In 300 to 500 years.
C. In 3,000 to 5,000 years.
D. In 30,000 to 50,000 years.

A, in three to five years. Greenland ice, built up over 15,000 years, shows that North America's average air temperatures can careen 20 degrees Fahrenheit from glacial to balmy within a few years. Between 10,000 and 15,000 years ago, Greenland's climate flip-flopped twice in an astonishingly swift three to five years.

Counting annual layers like tree rings back 15,000 years, scientists distinguish winter from summer ice by differences in atmospheric dust and snowfall. Studies of Barbados corals, Swiss and Polish lake sediments, and Antarctica, New Zealand, and the South China Sea suggest that abrupt climate shifts had worldwide impact.

►185 What makes climate switch fast from warm to cold?

The sudden sinking of the North Atlantic's surface water. During the ensuing ice age, the ocean's circulation would stagnate. Layers would form with fresher, warmer waters on top, and saltier and colder (and hence denser) water below. Gradually, over hundreds or thousands of years, some heat would diffuse down and some salt would diffuse up. Over time, the density of the bottom layer would lessen. Finally, some surface water could pierce the bottom layer, suddenly jump-starting the ocean's circulation again.

►186 More than half the Atlantic Ocean's volume comes from two sources. What are they?

Cold deep waters from the North Atlantic and from the Weddell Sea off Antarctica. Roughly one million cubic yards per second form and flow from these sources. North Atlantic deep water is relatively warm and salty: 36.5°F and 3.5 percent salt. Its salinity is the result of moist, low-level winds that export water vapor from the Atlantic across Central America into the Pacific. Weddell Sea deep water is colder (2°F), fresher (3.46 percent salt), and denser; thus it flows along the ocean's floor.

Global warming could cause enough precipitation to lower the Atlantic's salinity and prevent the Atlantic's water from over-turning.

▶**187** **How fast does the Atlantic Ocean completely overturn its water, bringing the dark, near-freezing depths up to the light of day?**

A. Every 6 years.
B. Every 60 years.
C. Every 600 years.
D. Every 6,000 years.

C, every 600 years. The cycle is propelled by differences in density between the North Atlantic's deep salty water and Antarctica's deep water, which is fresher, colder, and denser. As these cold waters sink, they release as much heat as the sun provides those regions in winter. A current transports North Atlantic deep water south along the coasts of North and South America. Winds blowing over Antarctica's circumpolar current help overturn the ocean, too.

Deep-water currents do not form in the North Pacific; its water is so fresh that it does not sink more than a few hundred meters. As a result, water lying 1,000 to 3,500 meters down in the Indian and Pacific Oceans is the oldest deep water in the oceans.

▶**188** **What is earth's largest geographic feature?**

A 37,000-mile-long system of deep-sea, volcanic mountain ranges that winds around the planet like a baseball seam, covering 60 to 70 percent of earth's surface. These midocean ridges bisect the Atlantic, Indian, and Pacific Oceans where their ocean floors are pulling apart at roughly the speed of a growing fingernail. Most seafloor, volcanic activity occurs along the ridge. Molten rock surges up through the crack from earth's mantle below and spreads and hardens to form steep hills and valleys; lava later coats the hills. Oceanic islands such as Hawaii, the Azores, and Iceland erupt along particularly hot spots.

▶**189** A park ranger was almost asphyxiated in a small, snow-covered cabin in the Sierra Nevada of California. Similar incidents occurred in underground utility vaults. Then, after four years of drought, acres of trees died near a popular Mammoth Mountain ski resort. What almost felled both ranger and forest?

A. Drought.
B. Insects.
C. Carbon dioxide.
D. Helium.
E. Hydrogen sulfide.

C, invisible and odorless carbon dioxide (CO_2) was smothering the land. Approximately 1,200 tons of carbon dioxide a day seeped up from deep inside Mammoth Mountain and deprived tree roots of oxygen. In places, between 30 and 96 percent of the gases in the soil was carbon dioxide. Doses lethal to people built up in a small cabin and storage vault. The appearance of volcanic gas often signals the reactivation of dormant volcanoes. Mammoth Mountain, a young volcano, last erupted roughly 500 years ago.

CO_2 smothered almost 2,000 African villagers when it bubbled out of two lakes and rolled over their homes during the 1980s.

▶**190** Corncob pipes, Indian peace pipes, oil pipes, water pipes—why would you want a stone pipe 300 miles deep and only yards wide in places?

Stone pipes cough up diamonds. Natural shafts through the earth's crust and mantle, the pipes are ancient, between 45 million years and 1.6 billion years old. Diamonds, among the oldest minerals on earth, formed 100 to 300 miles underground up to 3.3 billion years ago under great heat and pressure. They can explode naturally up through stone pipes to earth's surface in a matter of hours. South Africa has more than 2,000 diamond pipes. Since the greatest mineral rush in Canadian history started in 1991, more than 100 diamond pipes have been found in the Northwest Territories. Most are under crater lakes formed by the volcanic debris that erupts with diamonds.

▶191 Hot rocks can be anything from stolen diamonds and heavy metal concerts to uranium ores and plutonium. Forget all those, and name the very hottest rocks on earth.

Komatiites, magnesium-rich lava that erupted at temperatures up to 1560°C—half again hotter than ordinary volcanic lava. Komatiite was discovered in South Africa in 1969 but has since been found in Canada, Zimbabwe, Brazil, India, Australia, and Finland. Although komatiite lava melts at superhigh temperatures, it is not highly viscous and forms large flows and pillowlike structures. Because komatiite is produced only when a high proportion of the mantle has melted, it provides clues to the composition of earth's interior.

About 10 percent of the world's nickel comes from komatiite lava flows in western Australia. Deposits formed when hot lava melted the rocks it was flowing over. Komatiite eruptions were quite common during the first 2 billion years of earth's history but have been very rare since. Earth's interior was hotter back then.

▶192 Where are deep earthquakes most likely to occur?

In the Tonga Trench near Fiji and the Samoa islands. Most earthquakes occur close to the surface, but about 40 percent originate 40 to 420 miles beneath earth's surface. Most of these deep earthquakes occur where one section of earth's crust is being forced under another section. Along the northernmost part of the Tonga Trench, the Pacific plate is sliding 9 inches yearly under the Australian plate—the fastest plate tectonic movement on earth. Significantly, deep earthquakes are particularly intense there.

Why all the excitement under Tonga? The subducting Pacific plate, well over 100 million years old, is much cooler than earth's other subducting slabs; therefore it can better store the energy that is released later during deep earthquakes.

▶**193** Fire and ice are supposed to be opposites. But at least one volcano lurks under Antarctica's polar ice cap. And now ice has been found associated with a volcano at the equator, too: Rabaul volcano, which erupted in Papua New Guinea in September, 1994. Where was Rabaul's ice?

Overhead, in its volcanic eruption cloud. The cloud contained more than 2 *million* tons of ice. It formed when seawater entered the vent of the erupting island volcano. Nearing the molten magma inside the volcano, the seawater vaporized and exploded with the volcano's ash 12 miles into the air. There the water cooled to −40°F and formed ice crystals around the ash and sulfur. Other sea-level or lakeside volcanoes may also form ice.

When Papua blew, some of the ash fallout was extremely wet, and both mud and salt rains fell. The good news is that ice crystals probably made Papua's ash and sulfur precipitate out of the world's air quickly. The bad news is that silicates in clouds have damaged and stopped airplane engines. Existing aviation instrumentation cannot distinguish between (1) harmless clouds of water vapor and ice and (2) volcanic ash clouds disguised by ice.

▶**194** The most catastrophic volcanic eruptions occur when molten rock near earth's core ascends in plumes through the planet's mantle and crust, erupts, and cools over hundreds of miles. The cooled lava is basalt, a common iron and magnesium rock; basaltic lava flows are called flood basalts. Where is the world's largest flood basalt?

A. Columbia River, northwestern United States.
B. Deccan Traps, west-central India.
C. Kerguelen–Broken Ridge, south-central Indian Ocean plateau.
D. North Atlantic, British Isles, Faeroe Islands, and Greenland.
E. Ontong Java Plateau, western Pacific.
F. The Siberian Traps.

E, the underwater Ontong Java Plateau. Its flood basalt covers nearly 750,000 square miles, an area two-thirds the size of Australia. Ontong basalt could bury the contiguous United States three miles deep. It was created 120 million years ago by a plume measuring 370 to 860 miles across. Fortunately, it erupted under water, so few species were extinguished. Over the next 3 million years, lava floods raised the global sea level 10 yards.

Lava floods can raise global temperatures and sea levels and cause acid rain, darkness, and other chemical and climatic changes. Plumes and lava floods are associated with environmental crises, like the one that wiped out dinosaurs.

Recently scientists learned that undersea flood basalts are similar to those above ground, such as the Kerguelen, Deccan, North Atlantic, and Columbia River. The Siberian Traps formed 248 million years ago when 97 percent of all animal and plant species on earth were eradicated.

▶**195** The Rocky Mountains have been rising for 40 million years. It's taken 70 million years to build the Hawaiian islands. The Deccan Traps are much larger; they covered a third of India with cooled volcanic lava up to 1.5 miles thick. How long did it take to build the Deccan Traps?

A. One million years.
B. 100 million years.
C. 1,000 million years.

A, one million years, a mere moment of geologic time. The Deccan Traps erupted about 65 million years ago when half of all species, including the dinosaurs, became extinct. Originally, the Deccan Traps were three times bigger; seafloor spreading has moved parts of them as far as the Seychelles Islands. Most flood basalts form in less than 4 million years.

▶**196** Where was the biggest deep earthquake?

400 miles below Bolivia's rain forests, on June 9, 1994. At 8.3 on the Richter scale, it was the largest earthquake anywhere in the previous 20 years. No one was killed because the epicenter was so deep. But something was gravely wounded—the scientific theory of deep earthquakes. Deep earthquakes occur where one dense oceanic plate collides with another and dives down into earth's mantle. Theoreti-

cally, they should occur about 250 miles down, where heat and shear forces rearrange the molecules of a material called olivine into denser spinel. The Bolivian quake, however, contradicted many elements of the theory. Whether its condition is terminal or not awaits further study.

▶197 At any one time, earth's core is cooling by roughly 30 trillion to 40 trillion watts of heat energy. How much of this heat escapes through deep-sea vents, hot springs that erupt through cracks and fissures in the seafloor?

A. One percent.
B. 5 to 10 percent.
C. 25 to 33 percent.

C, 25 to 33 percent. Once thought to be rare, these vents dot a 150-mile stretch of volcanic mountains under the mid Atlantic Ocean. Spaced roughly 20 miles apart, they are close enough for deep-water animals and bacteria to drift from one chemical-rich hot-water oasis to another. Based on the amount of energy that earth's core is losing, there could be 5,000 such deep-sea vents worldwide. Some of the vents produce mounds of sulfuric precipitates as big as the Houston Astrodome or a 15-story chimney.

▶198 Blind sea shrimp see. Discovered in 1989 clambering happily around seafloor geysers that spit superhot water, the shrimp appeared to be eyeless. Then sharp-eyed scientists spotted eyes facing up on shrimp backs. Why up? Solar light rays can't reach 2 miles down. What kind of light lets shrimp see in the dark?

A. Light waves, invisible to people, produced by geyser heat.
B. Crystalloluminescence, produced by chemicals crystallizing.
C. Sonoluminescence, produced by the sound of bubbles popping.
D. Triboluminescence, produced by rock crystals cracking.
E. Cerenkov radiation and scintillation, produced by decaying radioactive elements in the vent water.

It's a murky problem, and B, C, and D are all possible.

▶199 A lowly tube worm holds the record for growing startlingly faster than any other invertebrate in the ocean. Obviously, it must live in the most nurturing, hospitable part of the sea. Where is that?

Next to a deep-sea hot spring 1.5 miles below the surface of the Pacific Ocean. Sound cozy? Yet in less than two years, a giant tube worm colonized the spot and grew almost 5 feet long. It's warm, after all, with 86°F water rich in chemicals and bacteria. There the tube worm happily spawns and grows almost a yard per year—in the dark. Elsewhere it probably could not even manage 6 inches yearly. Water from deep-sea vents contains little oxygen but lots of hydrogen sulfide, which is usually highly toxic to life. Nevertheless, more than 300 new species—many of them quite large—live near vents. These poison-loving creatures may reveal ways to treat water pollution. Technically, the giant tube worm, *Riftia pachyptila,* is a species of vestimentiferan tube worm.

▶200 How far below ground can bacteria live?

At least 1,500 feet *under* the ocean floor. Bacteria have been found living in organic debris that settled to the Pacific floor more than 4 million years ago. Scientists had assumed that water pressure and heat from earth's core would extinguish all life that deep. Yet anaerobic bacteria have not only been found feeding on methane and other hydrocarbons far below the briny deep, but they have also been "caught in the act"—dividing and multiplying. The mass of all the organisms below the seafloors could outweigh all of life *on* earth.

▶201 Life has flourished in the oceans for 3.5 billion of earth's 4.6 billion years, but life on land is a more recent story. When did the world's oldest known land fossils thrive?

A. 500 million years ago.
B. 800 million years ago.
C. 1.2 billion years ago.

C, 1.2 billion years ago, much longer than expected. The fossils are tubular microorganisms one-tenth as thick as a human hair but up to 150 microns long. They lived in Arizona soil that has since become rock. The structures may have been made by fungi or by bacteria, including cyanobacteria. Until their discovery, the oldest known land fossils were a mere 500 million years old. Arizona's pipes have stoked a scientific pipedream: If food was available 1.2 billion years ago, can a micropredator be far behind?

Scientists searched for land fossils in Arizona and California because rocks there have the same signature ratio of carbon isotopes that develops in modern soils where photosynthesis occurs. Similar microfossils 800 million years old were found in California.

▶202 What's a megaberg?

A. A Whopper.
B. A big city, for example, The Big Apple.
C. A big mountain, such as Mount Everest.
D. A giant iceberg.

D, obviously a giant iceberg. Megabergs with keels deeper than any known in the open ocean today may have drifted out of the Arctic during the last Ice Age, which ended about 15,000 years ago. Long grooves discovered in the Arctic seafloor off Spitzbergen suggest that icebergs descended 2,000 feet or more below sea level, far deeper than previously thought. The thickest Antarctic icebergs today may be only 1,200 feet deep.

The absence of scars on the underwater Yermak Plateau suggests that a floating ice sheet 1,200 to 1,800 feet thick may have covered the Arctic Ocean during the last Ice Age. If so, it could have linked glacial sheets covering North America and northern Europe.

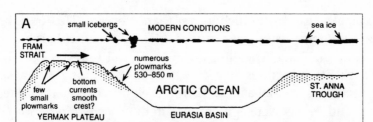

Deep grooves in sea mud were revealed by sonar images of the seafloor in the strait between Greenland and Norway that links the Atlantic and Arctic Oceans. Typical grooves are up to 5 meters deep.

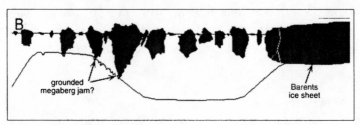

Icebergs probably plowed the grooves.

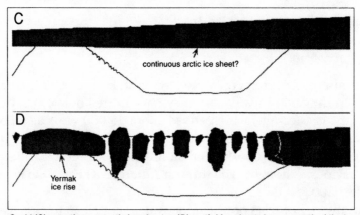

Could (C) a continuous arctic ice sheet or (D) partial ice sheets have smoothed the top of the undersea plateau?

▶**203** Mao Tse-tung is more famous for his Little Red Book than his Great Green Wall. The former aimed at mind control, the second at weather control. What was so green about Mao's Green Wall?

300 million trees since the 1950s in perhaps the biggest tree-planting program in history. They grow in the path of the Great Wall and prevent dust from China's arid north from moving south.

At one time, 10 to 20 springtime dust storms raged for days at a time and reduced visibility in Beijing to half a mile. Today, dust blows but rarely reaches storm proportions.

▶**204** Permafrost—soil, sediment, or rock frozen year-round—covers about 19 percent of earth's land surface and can be a mile deep. But in summertime, permafrost can get hot to the touch. So what accounts for the "perm" in permafrost?

Sweat and a thick mat of insulating sphagnum moss. Under the moss, the top layer of permafrost can melt on a hot day but keep relatively cool through evaporation. In human terms, it sweats. Eighteen inches down, the permafrost does not even know it's summer. By remaining permanently frozen, permafrost locks in vast amounts of methane gas, a potent greenhouse gas implicated in global warming.

▶**205** When Archaea bacteria were discovered in 1977, they were regarded as weird exotics thriving only in deep-sea vents, salt flats, and acidic springs with hotter-than-boiling water. Now Archaea have been discovered in frigid ocean water. Just how common is exotic Archaea?

A. As common as 750 elephants in a square mile of African savanna.
B. As common as an ice field under a volcano.
C. One of the most abundant organisms on earth.

Certainly A and possibly C. In frigid surface seawater off Alaska and Antarctica, Archaea compose 30 percent of all plankton less than one millionth of a meter across. Archaea thrive in deeper water, too. So Archaea may be bizarre, but they are as common as—well, as common as elephants used to be in the African savanna. Other new microorganisms are waiting to be discovered in plankton, too.

Like bacteria, the cells of Archaea have no nuclei. Nevertheless, Archaea are genetically more akin to people than bacteria. In fact, Archaea and bacteria are as genetically different from each other as any organisms known. Today Archaea are regarded as the third major evolutionary limb on the tree of life; the other limbs are (1) organisms with nuclei in their cells and (2) those without nuclei.

▶**206** To northern motorists, black ice spells danger because it forms an invisible but slippery coating on roads. At the North and South

Poles, dark brown and almost black ice forms a safe winter haven. What does black ice protect at the poles?

More than 100 algae species and many other microorganisms that overwinter in tiny, brine-filled channels deep within the polar ice. Only pure water freezes, so sea salt concentrates in cracks. Inside them hide single-celled algae, bacteria, fungi, protozoans, and animals such as amphipods, copepods, and marine flatworms. Ice dwellers cope with darkness, below-freezing temperatures, and brine seven times saltier than seawater. To survive, they slow their metabolism, enclose themselves in thick-walled cysts, or make protein antifreeze that liquefies their immediate surroundings. In the Antarctic, tasty algae grow on ice floe bottoms all winter, a veritable salad bar for hungry krill.

In spring, when light returns, the algae explode into growth and tint the ice dark brown and almost black. As the dark ice melts and microorganisms return to the open ocean, some give off a form of bromine that helps destroy the ozone layer.

▶**207** Name a 4,000-mile river approximately 150 miles wide and one mile deep that carries as much water as the Amazon?

A river in the sky. A wispy network of immense, water-vapor rivers—roughly five per hemisphere—flows one to two miles above the planet. A typical sky river could begin in the North Atlantic or northeast Pacific Ocean and waft to Greenland. Atmospheric rivers may form when water from equatorial oceans evaporates, rises, and drifts north or south toward the poles. Forming vast networks, they feed cyclones and hurricanes. Thus sky rivers feed rainwater to land rivers below.

▶**208** Elves and sprites dance through the air, but never with Santa's reindeer. If they have nothing to do with Donner and Blitzen, what do they consort with?

Lightning. There's a lot more to lightning than meets the eye. Elves, red sprites, and blue jets are new forms of electric fireworks discovered during the 1990s.

The ELVE blazes highest in the sky. ELVE—short for "emissions of light and very low frequency perturbations due to electromagnetic pulse sources"—are blazing disks 250 miles wide at the edge of space 60 miles up. The disk flashes on and off in less than one-thousandth of a second.

Red sprites look like blood red or pink jellyfish and sometimes sport red, blue, or purple tendrils. They can be 10 miles wide and 40 miles high. More common than ELVE, they generally occur about 60 miles up after an especially strong, positively charged, cloud-to-ground lightning bolt.

Blue jets spout from the tops of thunderheads and rise 20 miles. Some flare out at the top like a whale's spout.

These exotic lightning forms may affect global weather, ozone depletion, and storm formation; damage space shuttles and high-altitude spy planes; and share properties with nuclear explosions. And there are probably several more types waiting to be discovered.

Conventional lightning is a spark that occurs less than 10 miles above earth's surface; it can scintillate more than a second.

▶209 How many lightning bolts can flash per hour?

A. 10,000.
B. 50,000.
C. 100,000.
D. 200,000.

B. In the United States, a peak of 50,000 lightning strikes per hour has been recorded by the National Lightning Detection Network. But adding their return strokes—their flickering aftermaths—boosts the peak number of lightning events to 100,000 per hour. A summer day can experience more than 200,000 flashes in the United States, and electrical current is transferred from cloud to ground more than 70 million times annually, according to the national network. Worldwide more than 100 lightning events are estimated to occur per second.

Typical winter storms do not include much lightning. But 5,100 flashes per hour were recorded during a blizzard of snow, tornados, and heavy winds that killed 200 people on the East Coast in March 1993. In all, 59,000 flashes were recorded.

▶**210** In 1493, Pope Alexander VI drew a north-south line through the Atlantic Ocean just southwest of the Azores. All lands not already claimed by a Christian prince west of the line belonged to Spain, he declared. All those east of the line were Portuguese. Nature, however, draws a line through the Pacific Ocean. What is it?

A dark green line fringed with whitecaps. It floats like a deep-sea snake for hundreds of miles roughly halfway between Hawaii and Tahiti. Less than 6 miles wide, it lies along a front where cold and warm waters meet. The line—visible from satellites, planes, and ships—is marked by whitecaps and by dark green water thick as soup with buoyant, single-celled algae, a species of the photoplankton *Rhizosolenia*. Microorganisms line up to chow deep down in the nutrient-rich cold water. When full, they adjust their buoyancy and migrate up to bask in the sun. The line doesn't form every summer or autumn, but when it does, animals gather to feast like boozers at a bar.

▶**211** Without drastic cutbacks in fossil fuel burning, average temperatures could rise 3 to 6 degrees Fahrenheit by the year 2100. What region of the world is warming fastest?

A. Africa.
B. Australia/New Zealand.
C. Antarctica.
D. United States.
E. Pacific Ocean.

B. Since 1900, New Zealand's temperature has warmed an average of 1.3 degrees Fahrenheit, compared to global increases of only 0.8 degrees Fahrenheit. Because cloud cover is increasing in the southwest Pacific, nights warm more than days. Scientists down under argue that the area—relatively free of urbanization and air pollution—is a good place to monitor climatic changes that occur independent of human activities.

►212 The Yukon River discharges more than 60 million tons of sediment into the Bering Sea each year. But pits and furrows dump even more. What makes pits and furrows?

Whales and walruses. California gray whales make pits, and Pacific walruses dig furrows. And between them, they turn the seafloor into a battlefield. The animals forage in shallow water between Siberia and Alaska for shrimplike crustaceans and clams, respectively. About 200,000 walruses live year-round in the Bering and Chukchi seas; about 16,000 California grays breed in Baja California and travel 5,000 miles north to feed in summertime.

The whale rolls on its side, suctions in sediment, sieves food through its balleen, and expels the unwanted sand and gravel. Its pits average 2.5 yards long, 1.5 yards wide, and 2.5 inches deep.

Walruses swimming through clam beds make sinuous furrows the width of their snouts and up to 150 yards long. They apparently spit out a jet of water to excavate the clams. Then the walruses suck the bivalve from its shell. Their cultivation seems to enhance the area's productivity.

►213 What's the dustiest place in the United States? *Hint:* It's not a place you'd normally want to avoid.

Those lush and tropical Virgin Islands that figure so heavily in the dreams of winter-weary northerners. Nevertheless, they are the dustiest place in the United States based on the number of fine, airborne particles smaller than 2.5 microns across. The Virgin Islands outflake the Grand Canyon, the Badlands, and Death Valley. How so? Dust from the Sahara Desert. Dust is as much a world traveler as any tourist. Chinese and Mongolian dust storms reach Hawaii about 20 times yearly, reducing visibility in even the world's most remote archipelago. African dust also contributes to summertime haze in the eastern United States.

►214 The Bay of Fundy in Canada has the highest tides in the world. They can rise more than a foot every 10 minutes. Although the difference between low and high water at Fundy is generally 40 feet, it can approach 55 feet. How many places in the world have substantial stretches of coastline with mean tides larger than 30 feet?

A. 15.
B. 150.
C. 1,500.
D. 2,500.

A dock at low tide in the Bay of Fundy.

A, approximately 15. They include areas along the west coast of Britain; near La Rance in Normandy, France; in southern Argentina between the Bahía Grande and the Gulf of San Jorge; near Cambay in the Gulf of Khambhat, India; and Kimberleys, Australia. The gravitational forces of the earth, sun, and moon generate the tides. But at Fundy the shoreline's shape, the water's depth, and the bay's width make water slosh in and out of the bay at the same frequency as the tides and amplify their effect.

▶**215** The gravitational attraction among earth, its sun, and its moon causes ocean tides. But it causes earth tides, too. What are they?

The ground beneath our feet rises and falls about 8 inches a day as earth's surface is alternately dilated and squeezed by gravity. During high tides, when the moon pulls the ocean toward it, earth is being stretched. During low, weak tides, earth is squeezed. Earth-deforming rhythms occur twice-daily, daily, fortnightly, semiannually, every 8.8 years, and every 18.6 years. The latter two cycles are caused by complicated interrelationships among the movements of the earth, sun, and moon. The 18.6-year tidal variation, for example, is caused by the 5-degree tilt of earth's orbit toward that of the moon. The timing of geyser eruptions in Yellowstone National Park is affected by these earth tides.

▶**216** What unusual location was popular for gardening in Mexico 2,000 years ago?

Floating gardens grown on reed and grass mats that were anchored to shallow lake bottoms. The soil was mud scooped from lake floors. Potatoes are still grown on such "floating gardens" in Lake Titicaca. A few floating gardens remain and are among Mexico's most fertile farms.

▶217 **Where did the end of Oahu, Hawaii, disappear?**

Into the Pacific Ocean. Today the island of Oahu comes to an abrupt stop at a mighty cliff, a tourist attraction called Nuuanu Pali. The original tip of the island plunged into the Pacific as a giant avalanche more than a million years ago. In 1991 sonar equipment mapped 1,800 cubic miles of rocky debris stretching more than 140 feet out from the island and covering more than 9,000 square miles. Some of the fragments are mammoth boulders 10 miles to a side. As the avalanche plunged into the ocean, it spawned a 1,200-foot-tall ocean wave.

▶218 **Like rivers cutting through land, ice streams are fast-flowing currents of ice that slide between banks of slower-moving ice. In western Antarctica, "Ice Stream B" stretches 30 miles across, a half mile deep, and 300 miles long. Yet despite its enormous size, it flows 6 feet a day between icy "riverbanks" that crawl 6 feet a *year*. What makes ice streams move faster than their icy banks?**

Ice streams slide swiftly along on a lubricating layer of soft and slurpy sediment. Volcanoes under the ice sheet may keep the bottom layer warm. In any event, the layer is waterlogged, as soft as toothpaste, a mix of rocks and small particles.

In contrast, glaciers flow between mountains, not ice. A slurpy lubricating layer has also been found under a fast-moving glacier in Iceland, however.

▶219 **"And Abraham . . . looked toward Sodom and Gomorrah, and toward all the land of the plain, and behold, and lo, the smoke of the country went up as the smoke of a furnace."**
—*Genesis 19:27–28*

What was "the smoke of the country"?

Probably the oldest historical record of an earthquake. Earthquake records near the Jericho fault in Israel extend back 10,000 years. Sodom and Gomorrah were destroyed in approximately 2000 B.C., and Jericho's walls came tumbling down about 1000 B.C. During earthquakes, southern California's dry ground also produces great clouds of dust, "the smoke of the country."

▶**220** The largest single outpouring of lava in historic times produced the Haze Famine, which killed 10,000 people, a quarter of their country's population. Where did the Haze Famine occur?

Iceland. An 8-mile stretch of ground split open in June 1783. Advancing 9 miles in a single day, more lava poured out than water flows over Niagara Falls. A thick bluish haze of sulfuric acid hung over Iceland that summer and fall, stunting growth and starving livestock. In Europe the following winter, Benjamin Franklin suggested that clouds from the eruption were causing unusually cold weather.

▶**221** One of the world's major rivers is one-third silt by weight. What's the dirty river?

China's Yangtze River, called the Yellow River. It carries 2 billion tons of sediment yearly—four times more than the Mississippi. The silt comes from deforested, yellow loess soil in the western uplands of China.

The Molecules

of Life

►222 How long is a giant sperm?

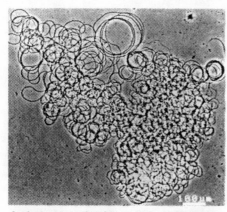

A giant sperm, that is, a single 58.29-mm-long spermatozoon obtained from the ejaculatory duct of a male Drosophila bifurca.

Roughly 1,000 times longer than human sperm and 20 times longer than the fruit fly that makes it. The giant sperm of fruit fly *Drosophila bifurca* is about 60 millimeters long. The fly's testes, where sperm is produced, account for 11 percent of the fly's body mass and a third of its abominal cavity. The testes are only a tad larger than the sperm.

In comparison, sperm of famous fruitfly *D. melanogaster* are a "mere" 1.76 milimeters long. Granted, that's 30 times longer than human sperm, but it's still tiny compared to its relative's.

Why make giant sperm? No one understands. To pass along as many genes as possible, males of most species produce enormous numbers of tiny sperm and mate as often as possible to distribute it widely. Fruit fly males produce giant sperm at great cost to themselves. Besides devoting 11 percent of their energy to sperm production, they mature later and produce fewer offspring per copulation than their more pedestrian cousins.

After all that work, most of a giant sperm never makes it inside an egg. Only about 1.6 millimeters of sperm tail enters *D. bifurca*'s egg. The rest—roughly 57 millimeters—is left outside, hanging. Perhaps giant sperm evolved in a race to reach and fertilize an egg. In contrast, the sperm of fruit fly *D. melanogaster* may be shorter but at least it gets all the way inside the egg.

►223 **How does the tiny fruit fly deliver its giant sperm?**

A. With a peashooter.
B. With a cannon.
C. With a remote-controlled missile system.

A suffices. Each sperm gets wrapped in a separate pellet and lined up in the male's genital tract. As the tract spirals down to a bottleneck, it forces each sperm to disentangle, roll into a compact ball, and wait in single file. In a 6-minute mating, the male's genital tract propels about 26 pellets one by one like peas from a peashooter into the female. The thrifty male reserves the rest of his sperm for other females. *D. bifurca* may be trying to monopolize the female's sperm storage chamber. If so, the female is one step ahead of him. She grows an enormous storage chamber—presumably big enough for his giants *and* those of other favored males.

►224 **Nile crocodiles can stay almost two hours underwater without surfacing for air. In fact, a crocodile often kills prey by holding its own breath *and* the prey underwater until the victim drowns. How doth the little crocodile manage such a breathless feat?**

Thanks to special hemoglobin, the oxygen-carrying molecule in red blood cells. As the crocodile holds its breath, carbon dioxide accumulates in its blood. The carbon dioxide, the normal by-product of respiration, drastically reduces the amount of oxygen that hemoglobin can carry. So the hemoglobin dumps its oxygen into the crocodile's tissues. Other diving mammals—including walruses, seals, and whales—store oxygen in muscle myoglobin, not in hemoglobin.

The sequence of amino acids differs greatly between human and crocodile hemoglobin. However, if only 12 amino acids in human hemoglobin were changed at key locations, people could have the oxygen-releasing abilities of the crocodile, too. Thus species adapting to new environments may not have to accumulate minor mutations over a long period of time; perhaps they can simply replace a few amino acids on a protein.

▶**225** Fingers and toes take a beating in the course of a lifetime. But the best-preserved DNA in a 2,000-year-old Egyptian mummy may be found in the skin of the fingers and toes. Why?

Enzymes make dying tissue digest itself. But the process, called autolysis, requires water. Skin and the peripheral parts of the body (e.g., fingers and toes) dry out relatively fast. After death, they may dry out faster than the body's DNA can be degraded.

▶**226** Name the species: About 10 new genetic diseases afflicting this species are discovered yearly. They include blindness, hemophilia, muscle and blood damage after exercise, heart damage, lameness, and so on. Its stillbirth and early death rates range between 15 and 30 percent and up. It may be more plagued with genetic problems than any other species on earth. What is it?

The dog. Of 500 dog breeds, most have one or more genetic diseases. Only 28 of the 170 breeds registered by the United Kingdom Kennel Club have no known genetic problems. For example, 30 percent of adult French briard herding dogs suffer from progressive blindness. Tragically, many Labrador retrievers bred to guide blind people suffered the same problem during the 1970s and 1980s. Dog breeding associations are financing research to identify the dog's genetic makeup, its genome. Genes produce an astonishing range of sizes and shapes in dogs, comparable to adult humans weighing anything from a few pounds to half a ton.

Many conditions caused by dominant genes have been eliminated; they carry disease risks even when inherited from one parent. Most remaining disorders are caused by recessive genes, which must be inherited from both parents before symptoms show up. Interbreeding of close relatives has been widespread, however. Inbreeding increases the chance of mating and reproducing with a partner who carries the same recessive genes. In comparison, mongrels have stillbirth and early death rates under 5 percent.

►227 Lightweight bones, aerodynamic feathers, highly efficient respiration—what else can a high-flying bird do to defy gravity?

Trim its DNA. Bird cells contain less DNA than the cells of mammals, reptiles, or amphibians. For example, chickens have shorter DNA regions of noncoding, nonsense genes than humans do. Slimming the DNA doesn't actually reduce a bird's weight, of course; but generally speaking, the shorter an animal's genome, the smaller its cells are. And because oxygen diffuses throughout small cells faster than through big ones, small-celled birds should be more metabolically efficient. Thus strong fliers (e.g., eagles and pigeons) have smaller genomes than weak fliers or flightless birds. Among mammals, bats also tend to have less DNA than earthbound creatures.

►228 In a student nightmare come true, Ruth was shocked to discover that she was enrolled in a music history course—with a big exam set for the following day. After staying up all night to outline a borrowed text, she got an A+ in the final with two points off for "style." But months later she was back where she'd started. She couldn't remember any music history. Why not?

The brain builds permanent memories only when it can rest between learning sessions. For example, fruit flies taught in one 20-minute block to associate a particular odor with an electric shock remembered the lesson three days. But fruit flies who rested 15 minutes between coaching sessions remembered for a week. Thus it appears that long-term memory involves a memory molecule called CREB, not CRAM. And unlike short- or medium-term memory, permanent memory requires the production of new proteins in the brain. Both fruit flies and mice use the same CREB protein to activate the gene sequences that produce the proteins. Further, different genes help produce different types of memory. Most important, mice who lack the memory molecule CREB forget their lessons. They might as well never have crammed.

▶229 Music lovers and parents take note: Infant baby songbirds actually listen to their fathers for about two weeks. Then young zebra finches practice singing, gradually reinforcing neural circuits in their brains as they memorize their fathers' song. All this time, the area of their brains involved in memorizing music is pumping a newly discovered protein named synelfin. After 35 days, as the birds approach adulthood, the synelfin suddenly starts to disappear. Why are gerontologists listening to baby birdsong?

A human form of synelfin has been linked twice-over to Alzheimer's disease, a brain disorder that destroys memory. First, people whose brains contain a protein shaped much like synelfin risk getting Alzheimer's. Second, fragments of human synelfin dot Alzheimer's dense brain plaques. Songbirds show, in simplified form, how the brain connects neurons to learn.

▶230 Because a female chimpanzee may mate with any and every adult male she meets, figuring out who actually fathered her young can be a hair-raising undertaking. DNA fingerprinting could assign paternity, but researchers did not want to risk tranquilizing tree-climbing chimps for blood samples. How did they solve this hairy problem?

They collect chimp hair from their sleeping nests. Naturally shed hairs with intact roots are especially useful because root tissue contains hair growth cells rich in DNA. Like people, chimps shed hundreds of hairs daily. Animal and human hairs, feathers, fish scales, plant matter, and beetles are all grist for DNA analysis. At digs and crime scenes today, sedimentary deposits are no longer thrown out; instead, they are searched for tiny organic remains that can undergo DNA analysis.

▶231 What makes new species: many small genetic mutations or a few big jumps? Darwin voted for the first view, saying, *"Natura non facit saltum"* (nature doesn't leap). Any problem with that?

This orange-red monkeyflower (left) is pollinated by hummingbirds; this pink monkeyflower (right) is pollinated by bees.

Monkeyflowers leap. When nature is under great pressure, evolution sometimes occurs in fits and starts. Just three key genetic changes turned an orange-red monkeyflower pollinated by hummingbirds into a pink flower that only a bee could love. Both *Mimulus cardinalis* and *Mimulus lewisii* grow in the Rocky Mountains and western United States, but everything else about them is different: size, color, shape, nectar amount and concentration, pollinator, and so on. Have molecular biologists spent too much time with fruit flies and not enough with other species? How would monkeyflowers vote?

▶232 Do you favor the flavor of rice or pasta? And why do you prefer one starch to another?

A. Your sweet tooth.
B. Your taste buds.
C. Your genes.
D. A craving for salt.

C, naturally. The starch you chew may depend on which genes you inherited for producing the starch-digesting enzyme, amylase. One variant of the gene sweetens starches differently from the other. For example, the shrimplike crustacean, *Gammarus palustris,* eats two types of algae in salt marshes along the eastern seaboard of the United States. Depending on its genes and the types of amylase they make, the crustacean prefers one algae to another. Humans also have different combinations of amylases in their saliva. Thus—in a crushing blow to gourmet cooks—the delectability of food may depend less on its inherent yumminess than on the diner's genes. Could the appeal of mothers' cooking be genetic?

▶233 Cells creep and crawl inside us all the time. Every day, 100 billion white blood cells called neutrophils extend their stiff, flat, gelatinous "feet" to glide out of a person's bone marrow. After several hours charging through the bloodstream fighting infections, these white knight cells settle down in other tissues. How fast can a cell crawl?

Each neutrophil crawls only about 1,800 millionths of a meter per hour. But so great are their numbers that the total distance traveled by all the neutrophils in a human body could take them twice around the earth each day. Selectively speeding or slowing the rate of cell crawl could spur healing, slow cancer growth, and fight infections. Controlling the filaments that stiffen the leading edge of a cell's gelatinous "foot" might also help thin the mucus that clogs the airways of cystic fibrosis patients.

▶234 How old is the oldest *living* organism on earth?

Between 25 million and 40 million years old. Spore of bacteria that old were resuscitated in 1995. The bacteria had been preserved in the abdomen of a 25-million-year-old stingless bee, preserved in turn in amber from the Dominican Republic. A close relative of the bacteria, *Bacillus sphaericus,* lives symbiotically in bees and other insects today. When food or water are scarce, some bacteria can form spores and grow protective protein shells around them. Inside their shells, the spores can dry out and remain dormant for years—that is, *millions* of years.

When spores form, two copies of DNA are made: One goes into

the spore, and the other may be released in the bee's gut. Only a bit of the bacteria's DNA remained in the 25-million-year-old bee, but when copied over and over, enough was produced to identify it.

▶235 **If the fittest survive, harmful mutations should die out, right?**

Not necessarily. Surprisingly, some actually spread. A mutant gene that causes a debilitating movement disorder appeared among the Ashkenazi Jews of northern Eastern Europe in the seventeenth century, roughly 12 generations ago. From one individual, idiopathic torsion dystonia (ITD) expanded 100 times so that today one in 1,000 to 3,000 Ashkenazi Jews carries the mutation.

There's another part of the ITD puzzle, too. Theoretically, a successful mutation should help those who inherit only one copy of it. People with one copy of the sickle cell anemia gene, for example, resist malaria better than those without; only those with two copies of the sickling gene risk serious illness. Yet the ITD gene helps no one; even those with one copy can get the disease. The mutation's success may result from a combination of factors: Perhaps it started with one wealthy Ashkenazi, whose many well-fed children survived, intermarried, and produced descendants marked by the rogue ITD gene.

▶236 **If the house mouse is trying to call your house his home, consider this. He traveled a long while and a long way to get there. Where did the original house mouse come from, and when did he start his journey?**

In northern India 900,000 years ago. House mouse *Mus musculus* is at once a world traveler and a homebody, who prefers to nest in human abodes. House mouse is the ancestor of all wild and laboratory mice, according to genetic studies of wild mice in India. Biologists there are developing a new strain of laboratory mice from wild mice caught in 30 different parts of their country. All laboratory mice today descend from three house mouse subspecies: *Mus musculus domesticus,* plus *Musculus musculus,* and *Mus musculus castameus.*

▶237 How do you tell one species from another?

A. Differences in form and structure?
B. An inability to interbreed?
C. Different genes?

The answer used to be A and B, but C may win yet. Chimpanzees in West Africa are a case in point. DNA fingerprints of their hair show that they have a different genetic makeup from two other chimpanzee species in East and central Africa. That unique characteristic could make western chimps a separate species. If true, they could have split from the others about 1.6 million years ago. The groups have few obvious differences, though only West African chimps use stone tools to crack nuts. (All chimps use twigs to fish for termites.)

▶238 What's the longest protein chain? And how many amino acids does it contain?

Titin, the longest discovered so far, is a muscle protein with roughly 30,000 amino acids. Proteins are long chains composed of small, amino acid molecules strung together. Protein chains stiff enough to keep their shapes have at least 30 or 40 amino acids. Shorter chains tend to flip and flop from one form to another. Many amino acids are organized in modules repeated over and over within chains. Mobile modules of up to 250 amino acids can move from one protein to another. Amazingly, they can even move from one distantly related species to another, for example, from animals to bacteria or viruses. Mobile modules, also called domains, may have played an important role in evolution.

▶239 The first living cell to get all its genes identified is guaranteed a place in the Science Hall of Fame. However, it plays an unpleasant role within the human body. What does it produce?

A. Dysentery.
B. Flu.

C. Earaches and meningitis.
D. Sore throats and cough.

B, *Haemophilus influenzae*. Despite its name, *H. influenzae* is no relation to the common flu. Roughly one millionth of a meter long, it lives only in humans. Its 1,743 genes, arranged in a circle, can be grouped into at least 16 categories, according to function, for example, replication, energy metabolism, and fatty acid metabolism. Inside the human body, the immune system and *H. influenzae* wage constant battle. During the struggle, *Haemophilus* changes its outer camouflage coat by varying its internal gene sequences, either by sloppy DNA replication or by acquiring new genes from fallen comrades. Surviving bacteria recognize the signature "bar code" embedded only in *Haemophilus* DNA and take up this molecule for gene-swapping or energy. The bacterium has turned a genetic weakness into a strength: It makes so many errors replicating itself that its genetic diversity is guaranteed, a big advantage in the war against the immune system.

▶240 What is the first complex organism to have its entire genome described?

Brewer's yeast, *Saccharomyces cerevisiae*. In possibly the largest biological collaboration ever, more than 400 scientists in 96 laboratories in Europe, North America, and Japan identified some 6,000 genes and 12.1 million base pairs for yeast. Yeast's genome is small compared to a human's 70,000 genes and 3 billion base pairs. Yeast is nonetheless the first eukaryote with a complete genetic map; the first genomes to be fully described were of bacteria. Scientists understand the biochemistry of yeast cells better than that of any other plant, animal, or fungal cell. But they still do not know what 40 percent of the identified genes do. Half, never seen before, have been named "orphan" cells because their families are unknown. Eukaryotes include plants, animals, and fungi—that is, any organism whose cells have nuclei.

▶241 Hundreds of patients have been treated with gene therapy since 1990. How many have been cured?

None, as of 1995. In fact, patients showed signs of improvement—mostly slight—in only 17 clinical trials. Cystic fibrosis and cancer patients barely benefited. Gene therapy's biggest successes are "bubble" children born with a defective gene so that they cannot make a vital part of the immune system, the enzyme adenosine deaminase (ADA). Without ADA, the children survive only inside sterile plastic bubbles. Given normal copies of the defective gene, the children can make it themselves. But supplementary injections of synthetic ADA may be helping the children more than gene therapy.

For gene therapy to be effective, normal genes must usually travel inside genetically altered viruses or liposomes to patients' cells so the missing proteins can be made. In trials, however, healthy genes reached only about one percent of their target cells and replicated inefficiently. The patients' immune systems also attacked the viruses. More knowledge about basic biology, immunology, and virology is needed.

▶242 Ancient DNA, even when badly degraded by time, can be retrieved and analyzed for clues to evolution and the past. Historical DNA was first retrieved from which of the following?

A. The 140-year-old skin of a quagga, an extinct horse relative, South Africa.
B. 4,400-year-old mummy skin, Egypt.
C. 7,000- and 8,000-year-old human brain tissue, Florida.

A, which showed in 1984 that the extinct quagga was more zebra than horse. But B and C were landmarks in molecular archaeology, too. B provided the first DNA from ancient human tissue; and C, taken from a bog, showed in 1988 that nuclear DNA from wet samples can be analyzed. In 1991, C also demonstrated that PCR, the polymerase chain reaction technique, can be applied to ancient human DNA. PCR copies short, damaged DNA fragments 25 to 40 times, doubling the amount of DNA each time, so that even DNA bits from a single cell can be identified. Molecular archaeologists study ancient biological matter.

►243 When did complex cells with nuclei first form?

Two billion years ago, according to a molecular clock timing the rate of protein change in different organisms. However, the molecular clock may be running a bit fast. Fossil records suggest that cells without nuclei had earth to themselves until about 1.4 billion years ago; only then did more complex cells with nuclei develop. If complex cells formed earlier, soon after life began on earth, odds are better that they also evolved on other planets.

►244 One of the world's richest troves of amber dates from the Cretaceous period, when dinosaurs lived and flowers started proliferating. Discovered in 1991, the amber dates from 90 million to 94 million years ago. Which of the following specimens embedded in fossilized tree sap are the oldest known?

A. Mosquito (with a mouth tough enough to bite a dinosaur).
B. Moth.
C. Biting black fly.
D. Mushroom.
E. Bee.
F. Terrestrial bird in North America.
G. Parasitoid wasp.
H. Ant.
I. Flowers in amber.

All of them. Cretaceous insects were beginning to use flowers as food, and flowers were beginning to use insects as pollinators and thus spread around the world. The fossilized flowers came from a primitive oak tree, while the moth's mouth suggests it was adapting to a diet of flower nectar. From 80 pounds of amber dug from deep clay mud, at least 100 new species were identified.

►245 Where was this amber bonanza found?

A. Latvia.
B. Lebanon.

C. Central New Jersey.

D. Costa Rica.

C, central New Jersey. It was discovered by an intrepid band of explorers, who bravely ventured forth from the American Museum of Natural History in New York City into the wilds of New Jersey for the sake of science. Analysis of the DNA in plants and animals trapped in the amber is expected to yield clues to evolution. Amber contains the best preserved protein and DNA on earth: Tree resin dehydrates any specimens trapped in it, and terpene chemicals in the resin preserve the specimens. Such a prize was surely worth a New Yorker's trip to New Jersey.

▶246 What and where was the first chicken domesticated?

A little red hen in Southeast Asia, according to genetic studies led by Akishinonamiya Fumihito, brother of the Emperor of Japan. A red jungle fowl subspecies, *Gallus gallus gallus,* came first—before the chickens or their eggs. Until recently, scientists thought that chickens were domesticated in the Indus Valley only 4,000 years ago. Then evidence from northern China's semiarid steppes pushed the date back to 8,000 years ago. The steppes are far from jungle fowl habitat, though. Genetic analysis of a trait that occurs worldwide in domesticated chickens, but not in quails or pheasants, located the little red hen.

▶247 Despite a grave shortage of fertile males, red fire ants from South America have spread through the American South. Fully 80 to 90 percent of male fire ants born in the United States are sterile. As a result, between 25 and 30 percent of all queen fire ants in the United States have no mates, compared with 1 to 10 percent in South America. What's the problem?

A. Diet.

B. Mineral deficiency in the soil.

C. The "founder effect."

D. Crowded habitat.

C, the "founder effect." Between 5 and 15 mated queens disembarked from a South American ship in Mobile, Alabama, sometime in the 1930s. These were the founders of *Solenopsis invicta*'s North American line. Each queen can produce approximately 240,000 workers. At least one of the founders had a genetic deficiency, however. Today, Argentinian ants have about 90 different versions of the sex-determining gene, while North American *invicta*s have only 10 to 13 versions. The fewer the versions, the more sterile males. Fire ants are a dramatic example of a "founder's effect," in which the limited gene pool of a small and isolated group of colonists can cause serious problems for their descendants.

▶248 How are puffer fish like salamanders?

Both suggest that junk DNA may serve no useful purpose. Puffer fishes survive quite nicely with a genome one-eighth the size of most invertebrates. Puffers have the same number of genes as other fish but fewer junk genetic sequences, which do not make proteins. Salamanders, on the other hand, are small creatures with enormous genomes, some 40 times larger than a human's. The size of their genomes varies radically from species to species. However, salamander species that have fast-replicating cells and that hatch earliest have the most streamlined genomes. Could less junk be more efficient?

▶249 If the nucleus of a cell were as large as a basketball, the DNA inside the nucleus would resemble fine fishing line. How long would the fishing line be?

A. 10 meters.
B. 200 meters.
C. 100 kilometers.
D. 200 kilometers.
E. 500 kilometers.

D, 200 kilometers, with its ends tied together. To reproduce itself, the DNA untangles, or as mathematicians say, it unknots.

▶250 This population became isolated in a mountainous home at the peak of the last Ice Age and remained genetically and linguistically untouched until recently. Their language is unrelated to any other spoken today. The group has also become genetically distinct with the world's highest frequency of the Rh-negative blood group gene and many distinct gene frequencies. Who are they?

The Basques of northern Spain and southwest France. They've been a distinct people there for at least 18,000 years, according to genetic studies.

▶251 What are the rabbit's closest relatives?

A. Ruminants.
B. Hyraxes.
C. Rodents.
D. Primates and tree shrews.

D, according to molecular studies of their proteins. The Old Testament believed in A, the Vulgate Bible in B, and zoologists have leaned toward C because of similarities in their skull, teeth, skeleton, and fetal membranes.

Chemistry

►252 The French are reliably reported to have 200 cheeses and one religion while Americans have 200 religions and only one cheese. But not even the French produce all the cheeses possible in this world. Which of the following animals produce milk that is made into cheese?

A. Buffalo.
B. Camel/dromedary.
C. Cow.
D. Goat.
E. Sheep.
F. Yak.
G. Zebra.

All but B. The first step in cheese-making is coagulating the milk protein casein. A female dromedary can produce up to 20 liters of milk a day, but her milk coagulates poorly because it is short on enzymes. As a result, the Saharan nomads of Sudan and Somalia—who own most of the world's 16 million dromedaries—must throw out whatever milk they cannot drink immediately. A portable cheese-making kit to set camel milk could make camelbert—as well as camelricotta, camelblue, and camelgouda. It's the taste treat we've been waiting for.

▶253 Saint Januarius was beheaded in A.D. 305, and more than a thousand years later his "blood" began liquefying and reclotting. Over the past 600 years, Saint Januarius's flowing blood has become history's best-documented paranormal phenomenon.

An archbishop frequently moves Saint Januarius's reliquary and its 30 millimeters of dark brown material during ceremonies held several times yearly in the Cathedral of Naples. Within minutes, hours, or days, the brown substance starts flowing freely. Returned to a vault for safekeeping, the material resolidifies.

How do scientists explain Saint Januarius's flowing blood?

Mud and thixotrophy, in which solid gels liquefy if stirred, shaken, or disturbed. Thixotropic gels—for example, ketchup, mayonnaise, and some paint and toothpaste—resolidify when left standing. A mudlike colloidal solution of clay would look and behave just like Saint Januarius's "blood." Fourteenth-century artists and alchemists could have used volcanic materials readily available from Mount Vesuvius near Naples.

The vital clue? The substance liquefies when moved for ceremonies or repairs, no matter the time of year or weather. Thus, motion is crucial, but temperature is not.

▶254 How does the black ink on this page resemble an iridescent opal?

Both are dehydrated colloids. In colloids, small solid particles are dispersed throughout a fluid. Ink, one of the earliest manufactured colloids, was used in ancient Egypt and China. Its carbon black particles are dispersed in fluid,

Polystyrene spheres suspended in ultrapure water form a lattice in this photograph of a colloidal system taken under a light microscope. The lattice in the image measures 30 by 24 micrometers.

applied to paper, and then dried, leaving the carbon black on the paper. Recycling the paper makes yet another colloid: oily ink droplets in pulp paper slurry.

Opals form from silica spheres dispersed throughout a fluid in a orderly, crystalline-like pattern. When the fluid evaporates, the crystallized silica is left behind. If the distance between the ordered planes of silica spheres is as large as the wavelength of visible light, they diffract the different colors of light and make the opal iridescent.

▶**255** How much methane fuel does a ton of garbage make?

A. 4 cubic meters.
B. 40 cubic meters.
C. 400 cubic meters.
D. 4,000 cubic meters.

C, 400 cubic meters of fuel over 10 to 15 years, although a landfill site will generate smaller amounts for 50 to 100 years. Over a decade, one ton of composting garbage can produce more than 100 times its volume in methane. Despite the potential, landfill operators do not maximize methane production.

As a greenhouse gas, methane may be 27 times more potent than carbon monoxide. Thus landfills are major contributors to the greenhouse effect. By burning their methane to produce electricity, communities could both reduce pollution and save fossil fuels.

▶**256** Who chawed history's oldest wad of chewing gum?

A teenager looking for a "high" in southern Sweden 9,000 years ago. An adolescent thoroughly masticated three lumps of birch resin gum and then—in a not unfamiliar move—dropped the gum on the floor of his (or her) hut. Imprints on one chunk come from adolescent teeth; that is, adult teeth, not yet worn down by Stone Age cuisine. The resin was toxic and unpleasant tasting, but its terpene compounds produced a buzz still enjoyed by some North American Indians.

Hunter-gatherers littered the Swedish hut with axes, dog droppings, thousands of hazelnuts, and bones of wild boar, deer, beaver, and small beluga whales. After the last Ice Age, rising sea levels and fine clay sealed off the wooden hut—and its yummy gummy resin.

▶257 **Name nature's greatest chemist.**

A. Butterflies.
B. Spiders.
C. Venomous vipers.
D. The tobacco plant.

A. Butterflies are leading contenders for the title. Each species of milkweed butterfly produces its own signature blend of 12 to 59 chemicals. African milkweed butterflies produce in their pheromone glands more than 200 chemicals, including hydrocarbons, alcohols, esters, and ketones—some never before found in nature. Many butterflies need an elaborate chemical-recognition system because they cannot rely on sight: to scare off predators, many butterfly species mimic the appearance of other, bad-tasting butterflies.

▶258 **Within minutes of a caterpillar attack, a tobacco plant counterattacks with the first of a cascade of chemicals. Within hours, the amount of poisonous nicotine in its remaining leaves has doubled. What does the plant do when a *mammal* bites?**

It *quadruples* its nicotine production. Even cutting the stem of a tobacco plant with scissors quadruples the poison in the plant's remaining leaves. In the process, a gram of tobacco leaves can accumulate as much nicotine as 100 unfiltered Camel cigarettes. Some species can even remember previous attacks: twice-bitten plants produced poison faster. Something in the oral secretions of caterpillars truncates this response, so that the plant can only double the amount of nicotine. Knowledge of how plants react under attack may help chemists develop ecologically benign pesticides.

►259 Modern lead pollution dates from the 1930s, when tetraethyl lead was added to gasoline to improve performance. But earlier civilizations also polluted with lead. Who fouled the air with enough lead to equal 15 percent of the pollution caused by twentieth-century gasoline?

The Greeks and Romans between 500 B.C. and A.D. 300. Roughly 400 tons of airborne lead reached Greenland during the 800 years of the Graeco-Roman era. Greeks mined lead sulfide for its silver content beginning about 500 B.C. Later, Romans sweetened their food with lead salts and mined lead itself for cisterns, roofs, pipes, and paints. Lead from their open-air furnaces belched into the air. By the birth of Christ, snow falling over Greenland was precipitating out about 2 picograms (2×10^{-12} grams) of lead for every gram of ice there. That's four times more than is deposited naturally by dust from the world's rocks and volcanoes.

By A.D. 1000, medieval Germans were mining and smelting silver and lead again. Again, lead levels rose until 1500 when approximately 4 picograms of lead settled on each gram of Greenland's ice.

►260 In autumn, when their larvae crave protein, the Japanese giant hornet specializes in mass, take-no-prisoners slaughter. Their intended victims? Honeybees. What happens if the hornet's well-laid plans go awry?

Japanese honeybees surround a giant hornet, at right.

Surprisingly, the hornet doesn't get stung. The bees bake it to death: 20 minutes at 116°F roasts the hornet without harming the heat-tolerant cooks. The Japanese honeybee *Apis cerana japonica* is one-twentieth as big as the giant hornet *Vespa mandarinia japonica*. But after decoding the hornet's pheromone signal, about 500 angry honeybee workers may race to their hive's entrance and surround an attacking hornet in a furiously vibrating ball of bees. Should their heated defense fail, adult honeybees can always escape and start a new nest.

Not surprisingly, Japanese bees don't share their roast recipe with introduced European honeybees. Finding a colony of European newcomers, a giant hornet marks the site with chemical pheromones to summon its pals. A gang of 20 hornets can kill 30,000 European honeybees in three hours and occupy the beehive for at least 10 days, feeding bee larvae and pupae to their own young.

▶261 **What unit do chemists use to measure energy?**

A. Joules.
B. Ergs.
C. Kilocalories per mole.
D. Kilojoules per mole.
E. Electron volts.
F. Wave numbers.
G. Hartrees.
H. Ridbergs.

A through H. Chemists are in a bind. According to the international system of standard units, energy should be measured in joules per mole. But they are far too big for atomic energies, which are typically about 10^{-19} joules. Physical chemists find electron volts more convenient.

▶262 **Like sparkling jewels, air bubbles catch and reflect light. But bubbles can also make light. In fact, bubbles can almost indefinitely generate flashes of light 30,000 times per second with clocklike regularity. How do air bubbles make light?**

By making light from sound. In sonoluminescence, an air bubble concentrates the energy of sound waves a trillion times and emits it again as light.

First, *ultrasound*—any sound pitched above human hearing—is beamed at a very small air bubble trapped in water. The walls of the bubble collapse at supersonic speeds, shrinking the bubble's volume a thousandfold. Gas compression and shock waves launched toward the center of the bubble heat its interior to 5,000 Kelvins at more than 2,000 atmospheres of pressure. The energy from the bubble's collapse breaks apart the molecules within the bubble. As the molecules recombine, they emit ultraviolet light. All this occurs so deep inside the bubble that its surface does not vaporize. And as ultrasound waves continue to strike, the bubble oscillates in size, emitting light flashes at every compression.

▶**263** Soil downwind from English crematoria is laced with a particularly poisonous element. What is it, and where does it come from?

Pure mercury—from dental fillings vaporized during cremations. Elemental mercury is lethal in even small amounts, but dentists have filled teeth with mercury alloys for 100 years. It is cheaper than gold and porcelain, and the mercury in alloys is not generally thought to be soluble enough to cause problems. Dentists in the United States fill more than 100 million cavities yearly, mostly with mercury amalgam. German health authorities, worried that small amounts of mercury could erode in the mouth and cause heart, thyroid, and menstrual problems, have banned mercury amalgams for pregnant women, small children, and kidney patients.

Between 1970 and 1990, the amount of mercury in the air over the Atlantic Ocean increased yearly from 0.16 percent to 1.46 percent—mostly from burning coal and waste.

▶**264** During three-and-a-half-hour marathon drinking sessions, the world's champion guzzler chugalugs 600 times its body mass in liquid—voiding continuously all the while. For its size, the half-inch-long male moth *Gluphisia septentrionis* imbibes as much puddle water as a 165-pound man who drinks 12,000 gallons at the rate of a gallon per second. What's the attraction?

Salt. The moth's larvae love poplar or quaking aspen leaves, a particularly poor source of sodium. Thanks to the male's drinking bouts, however, salt concentrates in its sperm packets for transfer to salt-deficient females during five-hour-long copulation sessions. The female then passes much of the vital nutrient on to her eggs.

Other adult moths and butterflies probably "puddle"—that is, sip at mud puddles—for salt, too.

The tiny male *Gluphisia* may be a champion voider, too. To avoid diluting the salty puddle it's drinking, it can jet its waste water roughly 20 inches away.

▶265 How might blue-gened blue jeans figure in your fashion future?

Many fabric dyes are hazardous and polluting, so blue jean makers hope genetic engineers can green the process, so to speak. Their goal? To insert indigo genes into cotton plants to make them grow blue cotton bolls. Indigo was the standard blue dye until the advent of cheap synthetic dyes. The first machine-spinnable, naturally colored cotton was grown in 1988. But so far, green, brown, gray, orange, yellow, and mauve cotton bolls have been easier to grow than true blue-jean blue.

Bioengineered cotton plants can make wrinkle-resistant polyester blends, too. The plant converts simple carbon compounds into linked esters—essentially biological polyester. The result is cotton fiber with polyester granules in its core.

▶266 What percentage of cigarette smokers *continue* to smoke even after a cancerous lung is removed?

A. 15 percent.
B. 25 percent.
C. 50 percent.

C, 50 percent. Fewer than 10 percent of smokers quit per year, although approximately 80 percent want to stop. The three classic signs of chemical addiction are all exhibited by smokers:

1. They can't quit.
2. They need larger and larger doses to reap the same amount of pleasure.
3. Withdrawal produces clear symptoms, such as headache, constipation, insomnia, depression, inability to concentrate, and anxiety.

►267 Since 1986, sporadic reports of a mysterious affliction called "bubble hair" have surfaced in the medical literature. What is bubble hair?

A. Bubble-gum-tangled hair.
B. Bouffant hairdos, last popular in the 1960s.
C. Silliness, empty-headedness; stereotypical female behavior.
D. A hot head.

D, a hot head. Bubble hair—patches of coarse, kinky, brittle hairs filled with bubbles—is an intermediate form of combustion. Once thought to afflict genetically defective hair, bubbles can plague anyone with a hair-drier blowing hotter than 350°F.

In another condition, "uncombable hair syndrome," the hair suddenly tangles like felt. Damaged surface scales on the hairs stick out and lock the strands together. Most "feltings" follow vigorous shampoos.

►268 Traditionally, beer lovers both drank and ate their suds. What put a stop to this merry but beery diet?

Bakeries switched to factory-made baker's yeast. Until the mid-nineteenth century, bread bakers used yeast, or barm, called brewer's yeast, left over from brewing beer. Their customers could literally eat their suds. Bakeries switched to factory-made baker's yeast produced from molasses because it is more reliable. During World Wars I and II when food became scarce, many Germans ate brewer's yeast again. It was manufactured on a large scale during World War I and added to soup and sausage to help replace the 60 percent of foodstuffs that Germany had imported before the war.

►269 Why do many beer lovers start sipping only after the bubbles in their brew change from spheres to polyhedrons?

Because then each sip of beer contains more liquid than foam. Gas bubbles in water-based foams start as small, thick-walled spheres but soon degrade into large, many-sided shapes. Maintaining the large surface area of the bubble requires vast amounts of energy, so the foams are highly unstable. Gravity drains water from the sphere walls into drain channels between the bubbles. As their walls thin, bubbles grow closer together and begin to influence one another. Gas inside small bubbles is under great pressure, so it flows into larger bubbles. At the point when the walls of low-pressure bubbles cave in, the beer's spherical bubbles become polyhedrons.

▶**270** **What molecule is so hard to make that bunnies eat their feces to get it?**

Vitamin B_{12}, one of the essential pigments of life, along with haem, the red in blood cells, and chlorophyll, the green in plants. B_{12}'s pigment is deep red. Dorothy Hodgkin discovered the arrangement of B_{12}'s 181 atoms in 1956, but how organisms make the vitamin in nature is still not completely understood. Only a few microorganisms make it, and many of them live symbiotically in the large intestines of animals. As a result, mammals get their daily B_{12} requirements from their intestines: rabbits by eating their feces, and people by absorbing it from the large intestine. The human system is not foolproof; we are the only animal species known to get pernicious anemia, the often fatal inability to absorb B_{12}. The pigment helps the body metabolize amino acids, the building blocks of proteins. It is also used to boost meat production in poultry and cattle.

▶**271** **A war was on. Afraid of a new German weapon, the U.S. government launched a crash, top-secret program to bring its scientists and engineers up to date. Name the weapon.**

Helium-filled blimps, called dirigibles, during World War I. Originally, Germany filled them with hydrogen to bomb London from altitudes of 16,000 feet. After British projectiles exploded the highly inflammable hydrogen, the Germans switched to helium. The second lightest element, helium is also the most chemically inert and does not burn. In July 1917, the United States funded a program to produce helium from natural gas. Production began too late in the war to affect its outcome, however.

▶**272** **Why is tenderloin more tender than stewing beef, even when it comes from the same steer?**

Because tenderloin has less collagen, or connective tissue. Different muscles of the same animal have different amounts of collagen. The more tender the cut, the less collagen. Tenderloin, for example, has one-third less collagen than stewing beef.

In general, heating meat tenderizes it by weakening the hydrogen bonds that hold the collagen together. Heat actually toughens tenderloin, however. It breaks the hydrogen bonds of the contractile cell proteins, the muscle parts that contract. As their protein chains unfold, they get tangled up with one another and toughen the meat. Tangled proteins also toughen cooked egg white and fried liver. That's why tenderloin is cooked rare: only enough to kill bacteria and flavor the meat's surface.

▶273 Roses are red, violets are blue—but not for long. What discovery may give red roses the blues?

The fact that flower cell sap controls the acidity—and hence the color—of flower petals. The pigment anthocyanin, which colors many flowers, changes like litmus paper in the presence of acids or bases. As the morning glory *Ipomoea tricolor* opens, for example, its sap grows more alkaline and its hue changes from purplish red to sky blue. Metal ions then interact with the pigment to stabilize the blue. We've always wanted blue roses, haven't we?

▶274 What plant makes plastic instead of starch? *Hint:* It is the first plant genetically engineered to produce a nonprotein.

The mustard plant *Arabidopsis thaliana*. It can be genetically engineered to produce and store a readily biodegradable polyhydroxybutyrate plastic (PHB) in its cells. What's next? Potato plastic and sugar beet plastic, they say. Till then, just pass the mustard.

▶275 Since boiling sterilizes objects by killing bacteria, boiling kills bacteria. Right?

Not always. Some bacteria bask in temperatures above the boiling point of water. Many live in superheated hot springs bubbling up through holes in earth's floor two and three miles under the ocean.

The springs originate when water leaks down through cracks in the seafloor and is heated by earth's magma under high pressures that suppress boiling. Most life around these hot vents exists where hot and cold water mix at temperatures between 86°F and 104°F.

Heat-loving bacteria have been found on land, too: at geysers 167°F and higher in Yellowstone National Park and in geothermal power plants.

Enzymes that can take the heat would be useful in genetic engineering, the production of sugars and detergents, the desulfurization of coal, and elsewhere.

▶276 How do you store bacteria that love temperatures hotter than the boiling point of water?

On a shelf, at room temperature. They'll survive there for years, just waiting to be heated up.

▶277 The United States has the world's highest fire death rate. Which of the following cause most American fire fatalities?

A. Home furnaces.
B. Kitchen ranges.
C. Electrical wiring.
D. Upholstered furniture and bedding.
E. Open flames in fireplaces.
F. Cans of paint and solvent.

D. Upholstered furniture and bedding cause one-third of the 6,000 fire deaths and 20,000 severe injuries that occur yearly. Mathematical modeling indicates that a phenomenon called flashover is often the killer. One cigarette burning in an upholstered chair can heat a confined room so hot that smoke and gas can ignite and raise temperatures by more than 1,100°C. At that point, the entire area bursts instantly into flame in a flashover.

Fire fatalities dropped from about 10 per 100,000 people early in the twentieth century to about 2 per 100,000 a decade ago. There has been little improvement since. New flameproof materials would save lives.

▶**278** According to the laws of physics and Chemistry 101, the higher the heat, the bigger the size. So what's a material that bends the rules and *shrinks* as its temperature rises to 1,000 degrees?

A ceramic blend of tungsten, oxygen, and zirconium known as zirconium tungstate. Some other oxides shrink in one direction, but this ceramic slims in all directions. Its tungsten and zirconium atoms are bonded, not to each other but to an intermediate oxygen atom. As the crystal warms, the vibrating oxygen atom pulls the other atoms closer together. Overall, the ceramic shrinks evenly by 0.75 percent.

The ceramic's shrinking success may make it useful in electronic circuit boards, where heat can make components swell and break apart.

▶**279** Benjamin Franklin sold his publishing business and studied the physics of electricity. The Italian Renaissance painter Piero della Francesca wrote about perspective and solid geometry. And Goethe, the German poet, contributed to physics, geology, and especially botany. What Russian composer considered chemistry his most important work?

Aleksandr Borodin. The illegitimate son of a Georgian prince and an army doctor's wife, Borodin was a chemistry professor at the St. Petersburg Academy of Medicine. A leader in the campaign to allow Russian women to become physicians, he helped organize the Russian Empire's first medical school for women. The government closed the school in 1887 after it had graduated 700 women doctors. Bitterly disappointed, Borodin collapsed at a faculty ball and died at the age of 54. Borodin, one of Russia's major nineteenth-century composers, was a close associate of Russian composers Nikolay Rimsky-Korsakov and Modest Moussorgsky.

▶**280** Clad only in his underwear, a graduate student (who else?) sat under a mosquito net and waited as one mosquito after another was let in to bite him. Three-quarters of the mosquitoes—of the type that kills hundreds of thousands of people yearly by giving them malaria—made beelines for his feet. Why did the mosquitoes make hamburger of the grad student's feet? What do they like?

A. Hamburger.
B. Limburger.
C. Footburgers.
D. Toenails.

B and D or, to be more accurate, their aroma. After the student washed his smelly feet with disinfectant, *Anopheles gambiae* mosquitoes kept on biting him, but they no longer focused on his feet. Human toenail clippings stink, thanks to fatty acids produced by the bacterium *Brevibacterium epidermidis*. The toenail bacterium is closely related, chemically speaking, to *Brevibacterium linens,* which makes Dutch Limburger cheese reek like—you guessed it—smelly feet. And sure enough, a trap baited with parfum de Limburger cheese also proved wildly popular with the mosquitoes.

Armed with the knowledge that toes attract these mosquitoes, chemists may be able to trap the little killers with the scent.

▶281 The moose in Isle Royale National Park in Lake Superior browse on sodium-depleted plants. As a result, these large ruminating animals get only 10 percent of their sodium requirement from the island's vegetation. With no salted peanuts or Fritos to supplement their diet, what can a resident moose do?

Spend the summer eating submerged and floating aquatic plants. They are so rich in sodium—between 50 and 500 times saltier than Isle Royale's plants—that moose get their yearly dietary requirements in eight weeks. Summer's sodium is stored in the moose's rumen, a large storage chamber in the stomachs of ruminants such as deer, sheep, and cattle. Moose bone may also store some sodium.

When sodium-short sheep are offered a choice of beverage—water or a sodium bicarbonate solution—they slurp up enough bicarb to meet their dietary needs. A region of the brain responds to the concentration of salt in extracellular fluid and stimulates the animals' appetite for salt.

Genes do not explain all animal behavior. Some—like a moose's summer salad—is determined by its chemistry.

▶282 What are the two most lucrative commodities smuggled into the United States through Miami?

A. Cocaine.
B. Heroin.
C. Diamonds.
D. Refrigerants.

A, followed by D, specifically chlorofluorocarbons (CFCs). The United States stopped manufacturing CFCs, except for "essential" uses such as rocket motors, in January 1996. But with a demand of more than 20 million pounds annually for the nation's 100 million automotive air conditioners, temptation is great. The ozone-destroyer comes onto the black market from Russia and its former republics and, to a lesser extent, from China and India.

Astronomy

►283 What caused the first traffic jam on the information superhighway?

The crash of the comet Shoemaker-Levy 9's fragments into Jupiter in July 1994. Observatories on every continent and in orbit around earth monitored the collisions 24 hours a day. Astronomers communicated using electronic mail on the Internet, but sometimes they couldn't log on to one another's home pages. Thousands of amateur astronomers and nonastronomers wanted to watch the collisions, too. In fact, 2 million images were downloaded for private viewing within the first 10 days of the event.

►284 The medieval walled town of Nördlingen, Germany, nestles inside an impact crater formed 15 million years ago by an asteroid's crash to earth. The town's stone walls ring the crater's rim 15 miles around. Why is the town's church such a gem?

Its stone is filled with diamonds. Unlike most diamonds, these were not created deep inside earth's mantle under great and steady pressures. Instead, they formed in a mere instant on earth's surface—out of thin air, so to speak. In the shock of the asteroid's impact, nanosized diamonds condensed from hot vapor clouds. What had been graphite or amorphous carbon formed silicon carbon composites never before seen in nature. Other meteorite impacts have also produced nanosized diamonds—up to several parts per million.

Incidentally, Nördlingen's crater is one of a pair. Approximately 10 percent of earth's 200 craters were caused by double asteroids, rocky fragments that either touched one another in space or were bound together by mutual gravity into an orbital pair. Doublet asteroids were discovered in the early 1990s.

►285 Until 1967, time was measured in terms of earth's erratic spin. Then atomic clocks became the official time standard. Today, a second is the time it takes an atom of caesium-133 to emit 9,192,631,770 oscillations at a particular frequency of radiation. A worldwide network of 200 atomic clocks is now accurate to within 200 billionths of a second per year. What could keep time even more accurately—10 times more accurately, to be precise?

Millisecond pulsars. Although they measure less than six miles across, they have the mass of our sun and spin at ferocious speeds. PSR1937+21 spins once every 1.557806448819794 milliseconds. A pulsar's advantage is its long-term predictability. Pulsars do slow down, but in a way that can be measured and used in calculations. PSR1913+16, which rotates once every 59.029995271 milliseconds, also orbits around another star in an unusual manner predicted by Einstein's general theory of relativity. If every clock on earth stopped, we could use pulsars to reset them.

►286 How long would it take *Star Trek* captains Kirk, Picard, Sisko, and Janeway to visit every galaxy in the universe, assuming they located one per episode and produced seven installments a week?

34 million years—provided there are no reruns. That's because the Hubble Space Telescope focused long and deep enough on a tiny sample of sky in 1995 to increase the estimated number of galaxies in the universe to 50 billion. That's five times previous guesses. Since a typical galaxy has 50 billion to 100 billion stars, the number of stars probably totals 5,000,000,000,000,000,000,000.

A truly tireless traveler willing to devote one second per galaxy 24 hours a day for 365 days a year could zip through them in 1,585 years, five months, 28 days, 16 hours, and about 37 minutes.

The most distant known galaxy is 14 billion light-years away. It was formed only one billion years after the Big Bang created the universe.

►287 A mysterious flash—as bright as a sun or an atomic bomb—blazed high above the Indian Ocean for several seconds in 1980. For months, the Pentagon worried that nearby South Africa had tested a nuclear bomb. Had it?

No, the flash was a meteoroid exploding with the force of an atomic bomb. Secret military satellites high in earth's atmosphere detected 136 meteor explosions between 1975 and 1992. Earth is bombarded yearly by roughly 80 meteoroids exploding 17 miles or more above earth with atomic-bomb force, according to military data declassified in 1993. Meteoroids are small bodies that speed through space and strike earth's atmosphere; they include comets of ice and asteroids of rock, iron, or nickel. In any case, they are leftovers from the creation of the solar system. Most meteoroid explosions go unnoticed because they occur over uninhabited land or ocean or behind clouds.

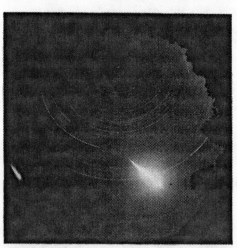

At 11:03 P.M. on May 5, 1991, a stony meteorite entered the atmosphere moving at 13 miles per second 55 miles above earth's surface. Eyewitnesses in the Czech Republic talked of "sudden daylight" and loud sonic booms. Within 4.5 seconds, the 30,000-pound meteorite had broken into bits 10 miles above earth. In this photo, a fish-eye lens opened all night revealed the meteorite's trail at center, star trails caused by earth's rotation, the rising moon at left, and a small semicircle, the Polar Star, near the North Pole.

▶288 Under ideal conditions, how many stars can a person see at night with the naked eye?

A. 2,000.
B. 3,000.
C. 6,000.
D. 10,000.

B. Ideally, roughly 3,000 should be visible in a mountain area far from any city in the Northern Hemisphere. Another 3,000 should be visible from a similar site in the Southern Hemisphere. But artificial light, a full moon, moonlight in general (which is really sunlight reflected from the moon's surface), smoke, dust, high humidity, and clouds make many stars invisible. Even in a dark suburb, only half as many stars are visible.

▶289 For the deadly serious drinker, where is the largest supply of alcohol to be found?

In space. The average molecular gas cloud contains more alcohol than people have ever managed to distill on earth. In fact, more chemistry occurs between the stars than on earth. Collecting enough alcohol for a drink may be difficult in space, though. Even a simple hydrogen atom may race a million years before combining with another.

Molecules form in dark environments whenever the temperature of matter cools below 3,000°C. Thus the first molecules were made about one million years after the moment of creation. Astrochemists study molecules in space.

▶290 How long does it take energy generated in the sun's core to reach the surface?

One million years. In the innermost 10 percent of the sun—a volume as large as Jupiter—temperatures reach 15 million Kelvins under such high pressure that gases are 10 times denser than gold or lead. There hydrogen nuclei fuse together to make helium and give off energy as X rays. Absorbed by nuclei and reradiated, an X ray can move about a centimeter in eight minutes—the same time that visible light waves take to travel 150 million kilometers from the sun's surface to earth.

▶291 What keen-eyed but hot-tempered astronomer carried a small box of salve in his pocket to reglue his gold and silver nose whenever it wobbled?

Tycho Brahe, a sixteenth-century Danish nobleman, lost his nose to another scientist in a duel over who was the better mathematician. Later, Brahe argued so bitterly with his royal Danish patrons that he moved to Prague with his priceless naked-eye measurements locating the positions of 777 stars. There, Johannes Kepler used Brahe's data to discover the principles of planetary motion. Incidentally, it is not known for certain whether Brahe lost the tip or the bridge of his nose.

▶292 Ironically, three of the greatest observational astronomers of all time—Tycho Brahe, Johannes Kepler, and Galileo—had vision problems. What were they?

Kepler was myopic and had double vision, possibly from astigmatism or misaligned eye muscles. Galileo went blind for a week after observing the sun through his telescope. An inadequate filter—perhaps smoked glass—may have burned his retina and left him temporarily snow-blind. Glaucoma later blinded his left eye and greatly impaired his right.

As for Brahe, his vision problems were meteorological, not medical. Brahe worked on a Danish island in the North Sea long before Galileo invented the telescope. Thus Brahe was trying to make visual observations through some extremely heavy cloud cover.

▶293 Although most mature stars orbit with one or two or even three companion stars, our sun is solitary. Will it ever find a buddy, or must stars acquire them at birth?

Our sun has lost its chance. Many astronomers once assumed that very young stars existed alone; then, when infrared radiation sensors were developed in the early 1990s, they discovered that young stars have even more companions than mature stars such as our

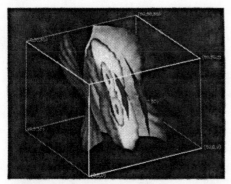

A binary protostar rests deep within a cocoon of the gas and dust from which it was formed.

sun. Among mature stars in the sun's neighborhood, only one-third are single while two-thirds have at least one companion star. Some almost touch their companions, while others are a third of a light-year apart. They may circle each other in hours or in tens of millions of years.

Star systems seem to originate in two violent steps over a few hundred thousand years. First, dense molecular hydrogen clouds collapse from their own gravity and become protostars. Second, the protostars break up into fragments. Football-shaped molecular clouds make binary star systems; pancake-shaped clouds make several-membered systems. The process is quick, considering that stars live several billion years.

▶294 The Human Genome project has grabbed headlines by cataloging the genes on our chromosomes. But molecular biologists aren't the only scientists using computers to compile encyclopedic collections of data. What are astronomers compiling?

The Sloan Digital Sky Survey, an imaging survey of about a quarter of the sky. It will automate the analysis, recognition, and classification of approximately 50 million galaxies, 70 million stars, and several million quasars. It will digitally record and catalog their images, colors, and positions (and for galaxies, their shapes and sizes) and will measure spectra of the brightest million galaxies and 100 quasars. The Sloan Survey uses an array of about 30 2-inch-square silicon chips. It detects objects about 20 times fainter than those found by the last large optical survey in the 1950s. That used 1,872, 14-inch-square, glass photographic plates.

In the meantime, Earth Observing System's satellite survey is studying our planet from above, and the National Biological Survey, assigned to compile an inventory of flora and fauna in the United States, has run into political trouble in Congress.

▶295   What's the world's largest telescope?

Planet earth, thanks to the Very Long Baseline Array of radio tele-scopes. Ten radio dishes, 25-meters-across, placed between the Virgin Islands and Hawaii turn a large part of earth into a giant radio-wave antenna. Headquartered in Socorro, New Mexico, the array produces details hundreds of times finer than anything observed by optical telescopes. It aims to measure continental drift on earth; material streaming out of quasar cores, where black holes are thought to be; and other objects in the universe.

▶296   Astronomers from the earliest shepherds to today—whether watching with naked eyes, optical glass, or radio antenna—had one thing in common. They observed electromagnetic radiation. It could be visible or invisible, ranging from ultraviolet to infrared rays, but it was always electro-magnetic. Now some astronomers are watching something new. What?

Neutrinos, the most numerous particle in the universe but the hardest to detect. Every second, nearly a billion billion neutrinos stream through each square yard of earth's surface. They bring clues to the brightest known objects in the universe (quasars and blazars) and the violence of black holes. But a neutrino passing through earth has only one chance in a billion of hitting something. Chargeless, virtually massless, and traveling virtually at the speed of light, neutrinos are extremely difficult to detect.

▶297   To study neutrinos streaming out of pulsars and galaxies around black holes, astronomers must filter out more pedestrian neutrinos created in the cosmic-ray showers over the North Pole. Where can astronomers go to escape them?

A. Lake Baikal, Siberia.
B. Peloponnesian peninsula, Greece.
C. Antarctica.
D. Cleveland, Ohio.
E. Pacific Ocean.
F. Mauna Kea, Hawaii.

A through E. Most neutrinos pass through earth, but a few interact with atoms to create muons. Muons live only a few millionths of a second, but they emit light flashes while passing through many miles of dense and transparent water or ice. Hence, neutrino hunters use earth as both filter and detector. For filters, they use salt mines in Cleveland and deep water in Lake Baikal, the Mediterranean, and the Pacific. Recently, they turned the South Polar ice cap into both filter and telescope. Using hot water, scientists bored holes 6 miles into the ice. At the bottom, pressure up to 200 atmospheres crushed air bubbles and made the ice as clear as glass. With the holes packed with detectors, the ice became a giant telescope. The Antarctic array can detect several hundred muon flashes yearly.

▶298 These beauties have enchanting names like Butterfly, Eskimo, and Helix. They come in every shape and form, too. Some are small, compact, and nearly round; others are bipolar, or double- and triple-shelled, or simply huge and shapeless. What are they?

Planetary nebulae. From earth, a planetary nebula resembles a glowing, fuzzy patch. At its center is a single, aging star spewing dusty gases into space. Fast winds force the gases into rings around the star. As the star continues to shrink, its surface temperature rises, producing so much ultraviolet light that it ionizes the ring of gas and lights it up as a planetary nebula. The luminous nebula spreads and thins for about 50,000 years, until it disappears from view. It has recycled much of the star's gas and in some instances fresh helium, nitrogen, carbon, and heavier elements back into interstellar space to form new stars—and eventually beautiful, new planetary nebulae.

▶299 White dwarfs—old, medium-mass stars that have cooled—shrink until they are as dense as 15 Cadillacs packed into a golf ball. That's an average density of 16 tons per cubic inch. After billions of years, the temperatures of some have dropped as low as 4,000 Kelvins. What happens to a white dwarf when it gets that cold?

Its gases start to crystalize. No white dwarf has lived long enough to cool to the point of invisibility, though. Our galaxy isn't old enough. All the white dwarfs ever born are still around to be seen.

▶300 What prominent member of the solar system is suspected of hiding a huge underground ocean of liquid ammonia 150 miles below its surface?

Titan, Saturn's giant moon. Titan's deep underground ocean may be a liquid mantle of mixed ammonia and water more than 100 miles thick. A second upper ocean of methane and hydrocarbons would be shallow and perhaps reach right up to Titan's surface.

Titan is one of the few parts of the solar system whose basic composition remains a mystery. That's because Titan hides behind a dense, orange haze produced from greenhouse gases. Titan is so large that it would be a planet if it orbited by itself. A moon 3,200 miles across, it is bigger than the planets Pluto and Mercury, though somewhat smaller than Mars. Titan and earth are the only solid bodies in the solar system with thick atmospheres, climate systems, *and* organic surface molecules.

▶301 A thin mixture of gases and dust fills the space between stars. Only heretics used to suggest that molecules exist there, but more than a hundred varieties of them have been identified. Some have as many as 13 atoms. Others are heavy elements, including arsenic, selenium, thallium, lead, tin, gallium, germanium, and krypton. What was the first amino acid discovered in space?

Glycine, the simplest building block of life. The protein-builder was found in 1994 in Sagittarius B2, a giant molecular cloud at the center of our galaxy.

Molecules help gas clouds evolve into stars. The molecules remove heat from the clouds' gas cores; when molecules collide with each other, some of their motion energy is transformed and expelled as light. As the gas clouds cool, they shrink under the force of their own gravity.

▶302 Venus has a nasty sulfuric acid atmosphere, but at least it doesn't have high winds. Right?

Hint: Earth's winds are fueled by differences in day-to-night surface temperatures and by earth's spin. In comparison, Venus has relatively constant temperatures and spins 1/243 slower.

Wrong. For some unknown reason, winds everywhere on Venus blow faster than the planet rotates. Near the cloud tops, Venus's winds blow three times faster than hurricanes on earth and 60 times faster than the surface of Venus rotates. Unfortunately, no more *Pioneer Venus* spacecraft are planned to answer the question. But solar heat absorbed near the cloud tops may cause the winds. If ever there was an ill-named planet, Venus is surely it.

►303 What is the most elongated object in the solar system?

1620 Geographos, a cigar-shaped asteroid 3.2 miles long and 1.2 mile wide that occasionally crosses earth's orbit. Ground-based radar imaged the asteroid in 1995, its closest encounter with earth in 200 years. Optical telescopes cannot focus on such small objects, and Geographos's portrait was one of the first of an earth-crossing asteroid.

Near-earth asteroids get renamed doomsday or killer rocks if they crash into a planet. One may have collided with earth 65 million years ago, when the dinosaurs died. Observation of near-earth asteroids could help explain how earth and other inner planets formed and how earth can be protected from future killer rocks. Most asteroids live—and stay—in a belt between Mars and Jupiter. They are thought to be rocky rubble left over from the creation of the solar system.

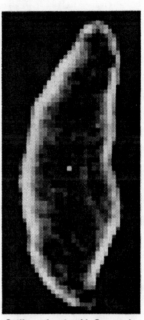

Outline of asteroid Geographos made from radar images obtained August 30, 1994. The asteroid was 4.5 million miles from earth, the closest in at least two centuries. The central white pixel is the asteroid's North Pole.

►304 Who was the most famous astronomer of all time?

Probably the most famous criminal mastermind of all time, astronomy Professor James Moriarty, in Arthur Conan Doyle's Sherlock

Holmes stories. Moriarty was modeled on Simon Newcomb, the most honored astronomer of the late nineteenth century. Newcomb was an intimidating Englishman, more feared than liked. He had attacked a questionable theory proposed by a friend of Doyle's who in turn became the model for Moriarty's evil chief of staff, Col. Sebastian Moran.

▶305 What's the faintest object that orbits a sun other than our own?

A cool brown dwarf. Called GL229B, it's located 19 light-years away in the Southern Hemisphere near Orion in the constellation Lepus. Brown dwarfs are cool objects midway in mass between planets and stars. For example, GL229B has 20 to 50 times more mass than Jupiter, but its core is still too small to get hot enough to sustain the nuclear fusion that can power stars for billions of years. As a result, brown dwarfs emit only one-hundredth as much radiation as the smallest known star.

Astronomers are sure that GL229B is a brown dwarf because its atmosphere contains methane, which exists only at relatively cool temperatures. Warmer brown dwarfs can be confused with faint stars. Theorists predicted the existence of brown dwarfs years ago, but finding one proved difficult. The difficulty stemmed from the fact that they are so dim and fade so quickly.

▶306 The supernova Crab Nebula is the remnant of a great stellar explosion. Who noticed the "most clear star" that became visible in the year 1054?

A. Medieval Europeans.
B. The Chinese.
C. The Arabs.
D. The Japanese.

A, B, C, and D. Politically correct scientists may be surprised. Many believe that medieval western Europe was not sophisticated enough to have observed a supernova, presumably because of Aristotle's philosophy and rigid thought control exercised by the Catholic Church. But Italians compiling the *Rampona* chronicle in Bologna reported in 1054 that, "At this time a most clear star appeared in the circuit of the first moon, beginning in the night of the 13 Kalends."

▶307 What's our nearest galaxy?

A. A dwarf galaxy in the Sagittarius constellation.
B. The Larger Magellanic Cloud.
C. Andromeda.

A, discovered in 1994. It's just over the back fence, so to speak: only 50,000 light-years from the Milky Way's center. The Larger Magellanic Cloud, formerly considered our nearest neighbor, is three times farther. Closeness does not always make good neighbors, though. Gravitational forces from the Milky Way have already stretched the dwarf's shape into a clumpy sort of dumbbell.

▶308 What glows brighter than a million suns—roughly 20 times a day?

A bursting pulsar, the brightest known source of highly energetic X rays, discovered in the middle of our Milky Way galaxy. It was also a new type of X-ray source. Most X-ray producers either flicker regularly or dazzle sporadically. But bursting pulsar GRO J1744–28 did both, a feat never before observed in one object. Because it flickered with X rays twice a second, it looked like a *neutron* star, a star that had collapsed into a sphere 6 miles across. Because it burst intensely every hour or so, it also appeared to be stealing mass from a bigger, ordinary, companion star. The two orbited each other every 12 days. The bursting pulsar was discovered with the Burst and Transient Source Experiment (BATSE) on board NASA's Compton Gamma Ray Observatory (GRO). GRO J1744–28 fizzled several months after its discovery.

▶**309** Besides earth, do any other planets orbit sunlike stars? Yes or no.

Yes. The first planet discovered orbiting an ordinary star like ours is only 42 light-years from our solar system near the Great Square of Pegasus. The planet is invisible because it hugs the glare of the star 51 Pegasi. The planet is almost as massive as Jupiter, however, so its gravitational pull makes 51 Pegasi wobble a tad—the telltale clue that tipped off astronomers to the planet's existence in 1995.

Earlier, other planets were discovered orbiting a pulsar, an extremely dense, fast-spinning star that emits radio waves.

▶**310** Oysters, crabs, shad, and terrapin are among the gastronomic wonders of the Chesapeake Bay, the largest Atlantic Ocean inlet along the East Coast. Now another natural wonder is having an impact on the meteoric rise—and fall—of the Chesapeake. What's the wonder?

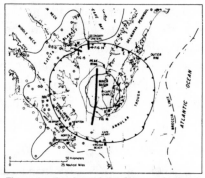

The crater that formed the Chesapeake Bay is the largest in the United States. It is 85 kilometers across and centered at Cape Charles, Virginia.

A meteor, of course. A meteor several kilometers across slammed into the southern end of the bay about 35 million years ago. It left an 85-kilometer-wide crater bigger than Rhode Island. It's the largest crater in the United States and the seventh largest known worldwide. It is half the size of the crater left by the impact that killed the dinosaurs. When the ocean rose millions of years later, Chesapeake Bay formed.

▶**311** Under a full moon, true lovers feel hearts warm, pulses quicken, and temperatures rise. At least one of these actually occurs, but which?

The third. During a full moon, earth's average global temperature is 0.03 degrees Fahrenheit warmer than under a new moon. Daily temperature data from satellites revealed the moon's influence over the temperature of the lower troposphere, the lowest 6 kilometers of our atmosphere. The lunar cycle also affects precipitation, thunderstorms, hurricanes, and cloudiness.

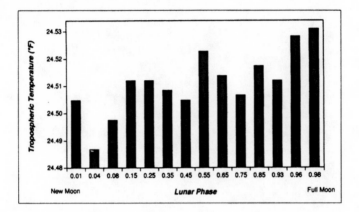

Physics

►312 The science fiction hero whips out his laser gun and either cools the hot-headed villain or fries him to a crisp. Which is it?

Could be either. Most lasers make heat, but shining a laser on some solids actually cools them off. If a laser of a specific wavelength shines on certain materials, atoms in the material absorb the light, combine it with their own thermal energy, and emit a higher energy photon. Having lost more energy than they had in the beginning, the atoms cool.

Coolers using glass doped with the element ytterbium could make compact optical coolers for electronics and detectors.

►313 How does a sand dune sing?

A. Like sexy sirens luring innocent caravaners to their deaths.
B. Like bells of buried monasteries.
C. Like the boom of a B-29.
D. Like the hum of a bee swarm.

A through D. Desert travelers have described the sound of dunes in all these ways. Dunes are noisy when they're squeaky clean and dust-free. Ordinarily, dust reduces the friction between sand grains. Without the dust, the friction between billions of moving grains sounds loud. Like shampooed hair, dunes are noisiest just after they've been washed.

▶314 Do hourglasses tick?

Yes, indeed. In art and grade-B films, hourglasses symbolize the inexorable march of time. Nothing, it seems, can stop those grains of sand from slipping slowly and steadily from the top of the glass to the bottom. Actually, however, hourglass sand stops and starts at regular intervals. Falling sand carries air from the top chamber of the glass into the bottom. Air pressure builds up below and begins pushing back up. The sand stops flowing, and the difference in air pressure—only about one ten-thousandth of an atmosphere—evens out. Then the flow resumes.

An hourglass starts to "tick" when its sand fills a certain proportion of the width of its neck: less than one-twelfth or more than one-half the width. Trapped air affects the flow of fine powders (e.g., cement and drugs). The flow of granular materials, which are neither liquid nor solid, is still poorly understood.

▶315 Driving through the black of a Canadian forest late one warm August night, a meteorologist swerved around a corner and suddenly it wasn't summer anymore. It was winter. The evergreen woods glistened brilliantly, as if covered with snow and drenched with the light of a full moon.

A. Was he drunk on moonshine?
B. Was he seeing ghosts?
C. Was he moonstruck?
D. Or was he blinded by sylvanshine?

D, he saw sylvanshine. As the meteorologist noted later, "This is not a subtle phenomenon." It occurs at night when an observer looks directly along a beam of light at certain waxy, dew-covered leaves. The wax covering the leaves or needles of some trees (e.g., juniper, cedar, blue spruce, hemlock, and rhododendron) forms spherical dewdrops. The spheres bounce so much of the light beam directly back at the observer that the leaves appear white. In the driver's case, his car headlights were reflected straight back toward him. Cat's eyes, bicycle reflectors, and highway markers also shine in the dark because they bounce light straight back instead of in many directions.

▶316 Women of the Kikuyu and Luo tribes in East Africa are fine, up-standing people—but they're freeloaders all the same. In fact, they sometimes outperform soldiers and experienced backpackers. What accounts for their prowess?

Whether carrying loads on their heads or not, the Kenyan women walk like pendulums. But loaded, they move like even better pendulums. They can carry heavy loads on their heads for free. That is, they can carry 20 percent of their weight on the tops of their heads or in slings hung from their foreheads without burning a single extra calorie. And they carry 70 percent of their weight more easily than army recruits and experienced hikers.

Perfect pendulums convert 100 percent of their energy back and forth between stored potential and kinetic energy. Because most people and animals convert only about 65 percent of that energy as they step, they make up the difference with muscle power.

▶317 Why is a ship's wake smooth? And why does it last for miles and hours behind the ship? Shouldn't the wake be *more* agitated than the surrounding water? It was, after all, churned up by the ship's hull and propellers.

The wake looks smooth because it lacks very short waves. A thin film of large molecules on the wake's surface damps low waves. The sea's surface often contains a thin film of organic molecules, probably from decomposed phytoplankton and other sea microorganisms. Ships moving through the film push it to each side like a snowplow piling snow on both sides of a road. Thick-edged "guard rails" form on each side of the wake and prevent short waves from crossing it, sometimes long after the ship has passed.

The ancients knew that pouring oil on troubled waters can calm a stormy sea. Today we know that a film one molecule thick can calm short waves. Without them, fewer long and high waves are generated.

▶**318** According to the Coriolis effect, bathtub water drains out in a clockwise spiral in one hemisphere and in a counterclockwise spiral in the other hemisphere. How twirls your whirlpool?

In both directions. As far as bathtubs are concerned, the Coriolis effect is actually the Coriolis myth. Fill a sink with water, stir clockwise, and let it rest 5 or 10 minutes. Then make the motion easier to watch by sprinkling the water surface with sawdust, flour, cornmeal, chopped whiskers from an electric razor, or the like. Then pull the plug and check the whirlpool's direction. Repeat the experiment, stirring the water in the opposite direction. Most sink water has some net angular momentum left over from its previous flow.

The Coriolis effect still applies to the rotation of weather systems around the globe but is too small to be observed in a bathroom.

▶**319** Crystallizing dust in the maelstrom of a plasma gas sounds as likely as freezing ice in hell. After all, plasmas are superhot gases of electrons and charged atoms moving at incredible speeds. Plasmas occur in hotspots like the sun, thermonuclear fusion, and space. And how many ice cubes exist in hell?

Apparently some, because dust can crystallize in plasma gas. In 1994 roughly 100,000 plastic spheres a millionth of a meter across were sprinkled on a plasma gas. They crystallized—at room temperature—into a disklike cloud visible to the naked eye. The 18-layered crystal hovered about for hours, although individual grains trembled from Brownian motion (particles in the plasma were bombarding them) and they spontaneously rearranged themselves a bit every few seconds.

At bottom is an actual image of the dust particles in the plasma. The inverted triangle at top is a schematic rendering to show the relationship between dust particles and plasma.

The plastic spheres crystallized because like-charged particles repel each other. Bombarded by electrons in the plasma, each plastic grain had acquired between

9,800 and 27,000 negative charges. Confined in a small space by electric fields, the particles could not move apart and adopted the most energy-efficient arrangement possible.

Plasma dusts occur in interplanetary space, Saturn's rings, proto-planetary clouds, and industrial plasma devices. At more pedestrian levels, they contaminate semiconductor manufacturing. They may help explain how frozen matter melts, a simple phenomenon not yet understood.

► **320** Ice skaters skate on water, not ice. That's how they overcome the friction that would normally stick their skate blades to the ice. So how does a thin film of water form between the ice and the blade?

A. The pressure of the blades melts the ice. When the pressure is released, the water refreezes.
B. All solids—even ice—are coated with liquid.
C. Heat from friction.

B is the leading contender. Friction (C) probably plays some role, too, but pressure (A) does not. The surface of crystalline solids is coated with a thin liquid film even when the bulk of the solid is too cold to melt. At freezing temperatures, the liquid surface of ice is about 40-billionths of a meter thick. Such surface liquids are often called quasi-liquids because their molecules are more ordered than those of a normal liquid but less ordered than those of a bulk solid.

As the solid gets colder, the liquid film thins. Thus at −35°C, the liquid film on ice would measure about 0.5-billionths of a meter. Not only would the skater get frostbitten, but her skate blades would stick.

Ice skating is still not completely understood. In 1842 the English physicist Michael Faraday first suggested that skaters skate on water. Advances in surface physics made during the late 1980s and early 1990s confirmed Faraday's inkling and improved catalytic converters and microelectronic devices.

► **321** This object is as dense as oak, but in 3 minutes flat, it can transform itself into a cup with streamers 65 feet long. What is it?

A modern parachute. It emerges from its loglike container in one second. Within two more seconds, it changes to a cup and streamers that can decelerate a one-ton payload from the speed of sound to highway speed limits.

A parachute, described by Leonardo da Vinci in the late 1400s, is still the cheapest and lightest method for decelerating objects in the atmosphere. Until recently, however, parachute design was more art than science. Modern chutes can rescue airplane crews, deliver military weapons, and recover the space shuttle's 175,000-pound solid-fuel rocket boosters.

▶322 Using atomic clocks, the 24 satellites of the Global Positioning System(GPS) orbit earth twice daily to transmit time and position data. Receivers listen to four or more satellites at once and measure the time difference between the transmission and reception of the satellites' signals. By calculating the distance between the receiver and each satellite, the receiver can determine its own location and accurate time anywhere on the face of the earth. How many atomic clocks does the Global Positioning System use?

96 airborne, including two rubidium clocks and two cesium clocks in each of the 24 satellites. Over the course of a day, they are accurate to within 10^{-8} seconds. Thus, over the course of one day—86,400 seconds—they drift by less than 10-billionths of a second. For military reasons, the U.S. government intentionally degrades the signal so that it is about 30 times worse. In 1996 the government decided to upgrade the signals over the next 10 years.

▶323 During the Persian Gulf war and the occupation of Haiti, American troops armed with Global Positioning System (GPS) devices always knew where they were. With too few military GPS units available for all the troops, soldiers had simply phoned mail-order catalogs stateside and used their charge cards to buy their own, commercially made GPS devices. The commercial units were quite accurate. During the emergencies, the U.S. Department of Defense turned off security measures that had made GPS inaccurate to within 100 yards, the length of a football field. And commercial manufacturers and scientists had long known how to circumvent the military's built-in inaccuracies. How accurate are the most precise GPS units available commercially?

To within a few millimeters, accurate enough for geologists to track continental drift and plate tectonics. Other nonmilitary uses include marine and aircraft navigation; routing of passenger cars and cargo containers; recalibrating offshore oil sites; and leisure sports. Navigational systems for the blind are still in the future.

▶**324** Which unit measures magnetic fields?

A. Tesla.
B. Gauss.
C. *B*-field.
D. *H*-field.
E. Oersted.
F. Amperes per meter.

A through F. All six measure magnetic fields. They aren't the only units either. Magnetism also abounds with henrys for inductance, webers for magnetic flux, joules per tesla squared per meter cubed, and 10^{-7} emu per cubic centimeter. In short, magnetism is a veritable Tower of Babel. Magnetism scientists simply didn't get the message when the Système International d'Unités (SI) recommended its worldwide set of standard metric units in 1960; but then, perhaps people who live in babelous towers can't understand messages. Newcomers to the field should not despair, however. Chemists have even more problems with their energy units.

▶**325** In a popular science demonstration, a small Pyrex beaker disappears inside a larger beaker. To do it, put a small beaker without any writing on it inside a larger beaker. Then fill the small, inner beaker with liquid until it overflows and surrounds the outside of the small container. If the liquid has the same index of refraction as Pyrex, the inner beaker "melts away." What liquid should you use?

A. A mix of 50/50 benzene and carbon tetrachloride.
B. Wesson oil.
C. Karo syrup.

B. Science professors sometimes use A, but it's carcinogenic. Wesson works wonders without the worry, despite a slight yellowish tinge. C doesn't work at all; light waves reflected from the interface between the glass and the Karo are emitted at different angles, so the line between beaker and syrup can be distinguished.

►326 Once upon a time—in 1897, to be precise—a state legislature tried to change the value of pi. A local physician had offered the legislators a deal they couldn't resist: In return for "legalizing" his new and improved value of pi, he would give the state free use of his copyrighted invention and collect royalties only from *other* states. Impressed with his largesse, legislative committees approved the new pi. Tipped off to the scam, local newspapers and a mathematician protested. Which state almost changed pi?

A. Arkansas.
B. Florida.
C. Indiana.
D. Louisiana.
E. Oklahoma.
F. Tennessee.

C, Indiana. The other states later tried to legislate another scientific principle, evolution. During the 1920s, Arkansas, Florida, Mississippi, Oklahoma, and Tennessee banned its teaching. Louisiana joined the pack in 1981. Arkansas's law, ruled unconstitutional in 1968, was reinstated in 1981.

►327 In 1995 the largest single numerical calculation in the history of computing targeted one of the following. Which one, though?

A. The value of pi.
B. Black holes.
C. Strings.
D. Glueballs.
E. Antimatter.

D. Gluons, which carry the forces binding quarks and antiquarks together inside protons and neutrons, can themselves get stuck together as glueballs, according to quantum chromodynamics (QCD).

At first, physicists weren't sure they could hope to find the tiny, short-lived glueball; their computers weren't powerful enough to answer the question. A specially designed and built computer with 566 processors, each a powerful computer in its own right, predicted that glueballs do live long enough to be detected. In fact, accelerator experiments may already have produced one.

▶**328** Rivers meander much the same way the world over. They don't lash wildly back and forth. Instead, they're generally mannerly and stay within well-defined strips of land. Why?

Rivers conserve energy by pinching off extreme bends, which then become U-shaped lakes. Exaggerated bends form when soil from the outside of a river's curve (where currents are faster) erodes and settles on the inside of subsequent bends. Eventually, it is more efficient for the river to flow straight by the loop, thus cutting it off. Understanding meandering may help geologists find oil; it forms when organic materials are compressed between sand deposits.

▶**329** The filaments in incandescent lightbulbs are made of tungsten because it has the highest melting point (3,653 kelvins) of any element. Unfortunately, the efficiency of a tungsten bulb is a mere 10 percent of its theoretical potential. What's the trouble with tungsten?

Like other objects at such high temperatures, tungsten emits between 80 and 90 percent of its energy as infrared radiation, which is invisible to the human eye. If Thomas Edison's lightbulb could be reengineered to emit *all* its energy as visible radiation, it would be more efficient than the best fluorescent lamp. Engineers hope high-performance laser mirrors will reflect the invisible infrared radiation back onto the tungsten filaments, heating the filaments even hotter and making them emit still more visible light; such a system would increase a lightbulb's efficiency by 25 percent. Over its lifetime, each bulb could save an estimated 125 pounds of coal.

Another scheme to reduce the amount of infrared radiation uses clusters of tungsten atoms as the light source; atomic clusters are too small to emit much infrared radiation.

▶**330** _____ is fickle, magnetically speaking. It is nonmagnetic in bulk quantities, but magnetic in tiny clusters of 15, 16, or 19 atoms. Larger clusters containing 60 or more atoms are not magnetic, though. So far, it is the only known material to behave in this odd manner. Fill in the blank with one of the following.

A. Platinum.
B. Rhodium.
C. Lanthanum.
D. Ytterbium.

B, rhodium, a hard, corrosion-resistant metal much like platinum. Some rare earth metals may be on-again-off-again magnets, too. Magnetic clusters could be useful in the recording industry.

▶**331** Michael Faraday, often called history's greatest experimental physicist, was the son of a poor blacksmith. There were weeks in Faraday's childhood when he had to make one loaf of bread last all week, and he managed to attend school for only one year. How did Faraday get the education to become a world-renowned physicist?

Apprenticed as a boy to a bookbinder and bookseller, Faraday learned physics by reading the _Encyclopaedia Britannica_ on the job. He studied chemistry by reading _Conversations on Chemistry_ by Mrs. Jane Marcet, an experimental physicist and popular science writer who collaborated with Benjamin Franklin on several experiments.

"When I questioned Mrs. Marcet's book by such little experiments as I could find means to perform and found it true to the facts as I could understand them, I felt I had got hold of an anchor in chemical knowledge and clung FAST to it," Faraday wrote. "Hence my deep veneration for Mrs. Marcet." Years later, Faraday and Mrs. Marcet became close friends.

▶**332** The traditional atomic nucleus is a liquid drop—specifically, a sharp-edged liquid drop that is extremely difficult to excite or break apart. Now some big exotic angels have overturned this familiar illustration. Who are these heavenly upstarts?

Nuclei with halos. Halos are wispy atmospheres of neutrons that surround nuclei much as electron clouds surround atoms. Compared to normal nuclei, halo nuclei are much larger and more fragile and often react strongly with other nuclei. Halos also disobey the laws of classical physics; they are strictly quantum phenomena.

The star of the halo nuclei is Li-11, a lithium isotope with eight neutrons and three protons. Two of its neutrons form a tenuous halo extending far beyond the nucleus. In laboratory experiments, Li-11 lives a few thousandths of a second.

In nature, halo nuclei exist in the center of the sun and in outer layers of neutron stars. Discovered in 1986, they may help explain how different combinations of neutrons and protons stick together as nuclei.

▶ **333** Many medieval cathedrals and monasteries were destroyed in France during the religious wars of the seventeenth century and the French Revolution of 1789. Identifying their remains—bits of limestone statues, columns, and capitals—stumped art historians. But recently they learned that two kings' heads in the Metropolitan Museum of Art in New York and the Louvre in Paris belonged to a French cathedral in Mantes. How did they match the kings with the cathedral?

By bombarding a matchtip's worth of powdered limestone from each sculpture with neutrons in a nuclear reactor. Twenty trace elements in limestone absorb the neutrons' energy, become radioactive, and give off the energy again as gamma rays. Every stone quarry produces a signature combination of gamma rays and trace elements, enabling historians to identify stone from 53 quarries in central France. The quarry database could be expanded worldwide.

▶ **334** Name the most sensitive detector known to science.

A. A SQUID.
B. An octopus.
C. A torsion balance.
D. An electrometer.

A, a SQUID, otherwise known as a Superconducting Quantum Interference Device. A typical SQUID can detect changes in magnetic flux corresponding to the energy required to raise one electron one millimeter in earth's gravitational field. The *best* SQUIDs are 100 times more sensitive.

SQUIDs are constructed like computer chips, etched and photolithographed films on silicon wafers. Because electrical currents in the human body produce many weak magnetic signals, SQUIDs can pinpoint brain lesions that cause focal epilepsy and areas in the heart that produce erratic heartbeats. SQUIDs also identify brain function around tumors, so surgeons can remove the tumor with minimal damage. Advances in high-temperature superconductors also permit SQUIDs to operate at warmer temperatures than before.

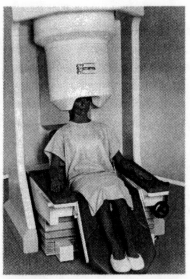

The SQUID consists of a square washer one millimeter across. The spiral coil on top connects to a superconducting coil to greatly enhance the SQUID's sensitivity to the magnetic field.

▶335 **What is the hardest substance known?**

A. Carbon nitride.
B. Silicon nitride.
C. Diamond.
D. Boron nitride.

A or C. Short chemical bonds make hard materials, and carbon nitride C_3N_4 may form a diamondlike lattice with even shorter bonds than diamond's. Unfortunately, physicists haven't made enough carbon nitride to be sure. If harder than diamond, carbon nitride would be useful in optics and industrial abrasives.

As for D, General Electric researchers years ago thought they had scratched diamond with boron nitride. The report soon appeared in the *Guinness Book of World Records*. Although it was soon realized that boron nitride does not out-diamond diamond, the false report remained in record books for years. Diamond, a latticelike crystal of pure carbon, has been the standard of hardness since Friedrich Mohs developed his hardness scale in 1822.

▶**336** The next time you get thirsty while shipwrecked on a floating Arctic ice floe, drink fresh water. But how? As far as the eye can see, you are surrounded by salt water and ice. Or are you?

You're not. Vintage ice is almost fresh water. Old pieces of ice have thawed and refrozen many times. In each cycle, the freshest water freezes first and leaves its salt concentrated in the surrounding water. This saltier water does not melt until the temperature gets even colder. After many years, the oldest ice has lost most of its salt.

By the same principle, frozen food contains quantities of unfrozen water. A 10 percent sugar solution stored at −10°C becomes pure ice—surrounded by a 56 percent sugar solution.

▶**337** After his students finished an experiment, the physicist wrote, "It was almost as incredible as if you fired a 15-inch shell at a piece of tissue paper and it came back and hit you." Who was the physicist, and what was the experiment?

Ernest Rutherford. Students of the British experimental physicist shot subatomic particles at extremely thin gold foil in 1909. Because atoms are primarily empty space, most of the particles went through the foil. Something heavy, however, bounced a few particles back, indicating that an atom has a nucleus. "It was then that I had the idea of an atom with a minute massive centre carrying a charge," Rutherford said.

▶**338** Some of the key discoveries in electricity and magnetism were made by amateurs. Who coined each of these electromagnetic terms and what were their full-time jobs?

A. Magnet.
B. Magnetic poles.
C. Electric.
D. Plus, minus, positive, negative charges.

A, Magnes, a Greek shepherd, who pastured his flock in a magnetic field. When the nails of his sandals and the tip of his staff stuck into the ground, he could hardly pull himself free. Digging, he found magnetite, a magnetic oxide of iron. According to Greek mythology, Magnes lived in Magnesia, a Greek province in Turkey, where magnetite was mined as early as 800 B.C.

B, Pierre de Maricourt, a French Crusader and military engineer, who wrote a friend about magnetism during a siege on August 12, 1269. De Maricourt placed magnetic needles on a natural magnet and watched them align themselves like the meridians of longitude on earth's surface. When the needles met where the North and South Poles would be on earth, he called their meeting points "magnetic poles." The Crusader realized that unlike poles attract each other and that each piece of a chopped-up magnet also has two poles. De Maricourt was one of the few experimental scientists during the Middle Ages.

C, William Gilbert of Colchester, physician-in-ordinary to Queen Elizabeth I and King James I of England, who coined the term *electric* for the force between two objects charged by friction. ("Electric" was Greek for amber, which people thought was the only substance affected by friction.)

Thanks to a queenly R&D budget, Gilbert proved that diamond, sapphire, amethyst, opal, carbuncle, jet, rock crystal, glass, sulfur, and sealing wax can become "electric," too—but that silver and copper cannot.

D, Benjamin Franklin, a publisher and diplomat who realized that there are two kinds of charge. He also knew that equal amounts of opposite charge neutralize each other. Franklin guessed wrong, however, about the direction of an electric current's flow: Electric current is actually carried in wires by negatively charged electrons moving toward positive charges.

Mathematics and

Computers

►339 Horses trot, canter, pace, and gallop. But their fancy footwork is nothing compared to the six-legged dance of the ancient and venerable cockroach. After all, a roach can do the tripod, the metachronal-wave, and—when really rushed—the four-legged or two-legged run. Why do cockroach gaits fascinate some mathematicians?

A cockroach moves like weights bobbing from interconnected springs. In mathematical terms, they're systems of coupled oscillators. Insect locomotion—like breathing, sleeping, and chewing—is rhythmic and symmetrical behavior controlled by a network of nerve cells dubbed the central pattern generator (CPG). Actually, its most important oscillations don't involve the legs, but the electrical circuitry that makes legs move. Understanding the relatively simple circuitry of the roach's nervous system may, in the short run (so to speak), help engineers build multilegged robots. In the long run, it may explain more complex, human systems.

In case you're curious, a tripod-ing roach moves three legs at the same time: the front and rear legs of one side step out with the middle leg of the opposite side to provide triangular support for medium and fast speeds. In the slower, metachronal-wave gait, three legs on one side move together in a wave from back to front;

then the opposite side takes over. The two- and four-legged quick-steps were discovered in American cockroaches in 1991.

▶340 How big a party must you throw before you automatically have either (1) at least three people who know one another or (2) at least three people who do *not* know one another?

A. 6.
B. 9.
C. 18.
D. 25.
E. 36.
F. 48.

A, a small one. Six people suffice, according to Ramsey theory, which ascertains how many of something must be present before a pattern of some sort appears. Ramsey theory suggests that complete disorder is impossible. Even when life seems most chaotic, some small segment of it has form and structure. Despite the theory, a small intimate dinner party may be in order for other reasons; as a matter of practical experience, large orgies tend to get unruly.

▶341 If you insist on throwing a bigger party, how many people must you invite to be sure that (1) at least four people know one another or (2) at least five people do not know one another?

A. 6.
B. 9.
C. 18.
D. 25.
E. 36.
F. 48

D, 25. To give you the answer to this vital question, 110 personal computers worked simultaneously, mostly in closed offices and laboratories late nights and over weekends. Eleven years of computer time were needed, one of the longest periods of computer time used for a purely mathematical problem as of 1993. Instead of checking 10^{65} possible combinations, researchers broke the problem into small parts that PCs could handle. Because of the enormous number of possibilities involved, mathematicians do not expect to solve larger Ramsey problems for many years. Ramsey theory was named for an English mathematician, Frank Plumpton Ramsey, who died in 1930 at age 26.

Four examples of parties with 24 guests, of whom at least four people know one another or at least five people do not know one another. A line joining two people shows they are acquainted.

►342 Friends meet for dinner at a restaurant that refuses to provide separate checks. Agreeing to split one bill evenly among themselves, the friends order their meals. No one chooses lobster, the most expensive item on the menu. Guests confine their booze to a glassful of wine apiece, and they all skip coffee and dessert. Are they:

A. Tightwads?
B. Dieters?
C. Close and chatty friends?

C. According to the theory of social dilemmas, they are probably a small group of friends who meet often and talk a lot. Otherwise, someone would run up a big bill, knowing others would split the tab. Small groups that communicate frequently are more apt to co-operate. After all, if they all retaliated at their next meeting by ordering lobster, everyone would owe big bucks. Social dilemma theory helps predict whether populations will conserve natural resources, recycle waste, donate to charity, reduce birth rates, or slow the arms race. In each case, an individual benefits in the short run by acting selfishly but cooperates for a long-term, common good. What a shame. Lobster's a lot more fun than pizza.

 One fine morning, the following mysterious events occur:

• Airplanes are grounded because they have not been overhauled in 99 years.
• Fresh bread is trashed, because it is 100 years old.
• According to your bank, the $1,000 you deposited last month has earned $400,000 in interest—almost a century's worth.

What's going on?

Computer chaos and algorithmic anarchy on January 1, 2000, as many computer programs erroneously treat 01-01-00 as the first day of 1900 instead of the first day of the new millennium. Many software programs record only the last two digits of any given year, assuming that the 1900s will last forever and forgetting that 1999 will eventually give way to 2000. Correcting the problem is difficult because old software has been modified many times and programmers often kept no record of the original program and changes. Clocks and dates deeply embedded in software programs regulate mortgage payments, interest rates, credit card numbers, insurance, airline reservations, pensions, loans, foreclosures, phone bills, and so on. A software engineering specialty, reverse engineering, hopes to automate the recovery of information from existing programs.

▶344 At 3 P.M. precisely on November 1, 1992, ATM customers trying to cash checks in New Zealand ran out of luck. Every Tandem CLX computer in the country stopped working. Tandems are packed with backup software and circuitry, so they are used in many automated bank machines and for electronic funds transfers. Two hours later, ATMs crashed in Australia. The shutdown swept around the world, from time zone to time zone. What had happened?

A glitch in the computer's internal clock cropped. up at 3 P.M. The Tandem episode is just a hint of computer problems expected to occur on January 1, 2000. There's another wrinkle, too: Only one out of four years that end in "00" is a leap year with a February 29, but 2000 is one of them.

▶345 What do the following phenomena have in common? A line of pelicans flies by, as synchronized as a Broadway chorus line. Armies of antibodies attack and wipe out invading bacteria. Some termites, master builders of the insect world, construct mounds up to a yard high and lace them with tunnels and chambers. Finally, a traffic jam.

All are complex patterns, but none is the result of careful planning by a central authority. Instead, each individual—whether pelican, antibody, termite, or motorist—independently follows a few simple rules. Birds and drivers match their velocities to those around them and keep safe distances from fellow travelers. The termite carries a wood chip until it finds another and drops its load; *or,* if it has no woodchip, it wanders around until it finds and picks one up. Eventually, the termite has dropped all the wood chips onto a few, large piles. Many such phenomena—including synchronized clapping at a concert and buying and selling on the stock market—are the result of decentralized decision making.

▶346 What do roaring earthquakes, Dow Jones crashes, and biological extinction events have in common?

They are rather unpredictable phenomena and are often attributed to large random events. But they may also be such unstable complex systems that the slightest disturbance can precipitate a disaster. Such systems have been identified in economics, earthquake studies, evolutionary biology's extinction events, solid-state physics, and astronomy. Mathematically speaking, they are examples of self-organized criticality.

▶347 How does a river flow toward an ocean?

Fractally and efficiently. It branches into smaller and smaller streams to save energy on its way to the sea. When physicists designed a computerized mathematical model of a river drainage system to minimize its expenditure of energy, the model was indistinguishable from digitized maps of real river basins. Now that river basins are known to be energy savers, scientists are asking whether other highly unstable complex systems are, too. During earthquakes, are local tectonic plates trying to minimize global stresses? Do trees arrange their branches so their leaves can get as much sunshine as possible? Stay tuned.

▶348 What book is considered the greatest and most influential scientific treatise ever written? It's been called one of the intellectual wonders of the world.

Isaac Newton's *Philosophiae Naturalis Principia Mathematica*—known for short as the *Principia*. The 1687 book underlies almost every aspect of modern science. In it, Newton introduced gravity as a universal force, explaining that the motion of a planet responds to gravitational forces in inverse proportion to the planet's mass. Using gravity, Newton explained tides and the motion of planets, moons, and comets. He also showed that spinning bodies such as earth are flattened at the poles. At the time, scientists believed that forces could be transmitted from one body to another only on contact; the idea of action at a distance seemed magical.

Newton and Leibniz independently invented differential calculus, but Newton first applied it to a major problem. He wrote *Principia* in 18 months, summarizing work that he had set aside 20 years earlier in order to pursue a potentially more profitable subject:

alchemy. *Principia* wasn't a big moneymaker either: Its first printing was only 500 copies.

▶349 **Barn owls look wise, but are they clever enough to help design a smaller, energy-saving analog chip?**

Perhaps. A barn owl pinpoints prey in the dark with a brain that operates like a parallel computer. It processes information from both sides of its head, dealing with slight differences in timing and intensity through separate circuits before it combines the information into a single message. A silicon computer chip—dubbed "Owl Chip"—reproduced in simplified form the owl's time difference analysis. But at 73 square millimeters, the owl chip was far bulkier and more energy-intensive than a real barn owl's far more complex nervous system. Biological principles—and owls—could help design smaller, more energy-saving computers.

▶350 **Thirteen centuries ago, approximately 40 Anglo-Saxons died, some abnormally, perhaps in a ritual. They were laid to rest at Sutton Hoo in England. Acid soil dissolved their flesh, bones, and clothes. All that remains are ghostlike impressions where their bodies compressed and discolored the sand below. How can archaeologists hope to bring to life the sand people of Sutton Hoo?**

By computerized measurements of their indentations in the sand. Archaeologists first scoop away soft sand, revealing discolored layers underneath. They encase any crumbling sand in rubber. Then, using a handheld probe, they make more than 3,000 measurements per body, graphing every centimeter of the surface. The computerized model is accurate to plus or minus 2 millimeters.

A computer rendition of a "sand man" buried in Sutton Hoo, England.

▶351 A celebrated French mathematician avoided colleagues for 31 years by pleading illness. Eventually, the editor of a mathematics journal discovered that the mathematician lived in a very famous establishment indeed. What was it, and why was the eminent scholar such a homebody?

A triple murderer, he was incarcerated for life in a large hospital for the criminally insane. André Bloch had murdered his aunt, uncle, and younger brother over lunch in 1917. Bloch, who had suffered a severe concussion during World War I, told his psychiatrist, "It's a matter of mathematical logic. There had been mental illness in my family. The destruction of the whole branch had to follow as a matter of course." A model prisoner, Bloch spent his time on mathematics and chess. Bloch was Jewish but escaped the Nazis during World War II by signing French aliases on his publications. Shortly before his death in 1948, the Académie des Sciences awarded him its distinguished Becquerel Prize. The triple murderer's mathematics is still regarded as "extremely pretty," "beautiful," and "sweet."

▶352 Please fill in the blanks.

- For every six larger, new software sytems put into operation, _____ others are scrapped. (Choose 0, 1, 2, 3, 4, 5, 6.)
- Developing the average software program takes _____ percent longer than anticipated. Large projects take even longer. (Choose 5, 10, 15, 25, 50.)
- _____ percent of all large systems are considered "operating failures" because they are not used or because they do not function as intended. (Choose 10, 15, 25, 50, 75.)
- _____ percent of American software programmers fail to keep track of the bugs they find, and the rest typically spot fewer than a third of the mistakes present. (Choose 30, 60, 90.)
- A typical innovation in software research is _____ years old before it becomes a standard programming technique. (Choose 1, 3, 6, 18.)

2, 50, 75, 90, and 18. As a result, some critics charge that much of software programming is a handcraft, not an engineering discpline. They complain about mass-market software makers who expect customers to report glitches—and buy corrections; other industries often guarantee mistake-free products. Critics charge that engineers rely more than software programmers on scientific theory, mathematical modeling, proven design solutions, quality control procedures, and handbooks of proven standards. Those who disagree with such complaints note that large software programs are among the most complex artifacts built by humans and that a failure rate of two out of eight means that six succeeded.

▶ **353** What standard unit of measurement rates the productivity of software programmers?

The number of lines of code written per worker per month. Unfortunately, this ignores differences in difficulty. As a result, some argue that computer science has no real way to gauge programmers' productivity. Fewer than 10 percent of American companies consistently measure their programmers' productivity. Is it any wonder?

▶ **354** Factoring large numbers is one of the most difficult computational jobs in mathematics. What is the best way to factor numbers?

A. Quadratic sieves.
B. Number field sieves.
C. Ramsey theory.
D. Long division.

D. Common, elementary-school-variety division is still the best way to find a number's factors. After all, half of all integers are divisible by 2, a third by 3, a fifth by 5, a seventh by 7, and so on through all the prime numbers (those numbers that are divisible only by themselves or by one). More than 95 percent of 100 randomly chosen numbers will have at least one factor under a million. Factors used in cryptography, however, are extremely large. Adding 10 more digits to the number being factored makes it 5 or 10 times harder to crack, so cryptographers often use 300-digit numbers. Current techniques cannot crack them.

▶ **355** To learn how far-flung computers can solve enormous scientific problems, 300 volunteers factored a 129-digit number using the Internet and 1,600 computers on five continents late nights and weekends for eight months in 1994. A mathematical technique called a quadratic sieve broke the 129-digit job into small bits suitable for PCs. Without the sieve, how long would the job—with shortcuts—have taken?

A. 40 quadrillion years.
B. 274 years.
C. 18 years.

A, 40 quadrillion years. What did team members get for their work? Decoding an encrypted message, they were rewarded with the statement: "The magic words are squeamish ossifrage." The words are nonsense: Ossifrages are bone-crushing vultures, so they are anything but squeamish. On the other hand, tackling a 40-quadrillion-year job is no task for the fainthearted either.

►356 What were the first nationwide communication systems to move detailed data hundreds of miles within minutes? *Clue:* They were invented before Morse's telegraph.

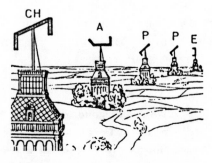

Semaphore towers of optical telegraph systems could transmit 20 symbols per minute across France early in the nineteenth century.

Optical telegraphs using either semaphore towers or mechanical shutters covered Europe 200 years ago. France and Sweden built the first systems in the 1790s. The French revolutionary government and Napoleon dubbed Claude Chappe's system of semaphore towers "tele-graph," meaning "far writing." Mechanical semaphores relayed numbers corresponding to 8,464 different letters, numerals, words, and phrases in a 92-page codebook. The system worked roughly as fast as early wire telegraphs. Thanks to data compaction, error recovery, flow control, and encryption, 120 stations across France could handle 20 characters a minute. By 1840 almost every European country had an optical telegraph system. They were gradually replaced by electromagnetic systems.

▶**357** What does horse racing have in common with rogue cops, Stone Age toolmakers, sex-befuddled students, twinkling stars, Shakespeare's writing style, Wall Street, and stroke victims who read "cat" for "cot" and "apricot" for "peach"?

Neural networks. These computer programs mimic collections of simple brain cells—hence the term *neural networks*. Like people, neural nets can be "trained" to find patterns within mounds of data and can draw conclusions based on incomplete information.

Neural nets identify the sex of people photographed without hair, jewelry, or makeup better than students can; read postal codes; predict stock and horse race winners; identify Chicago cops who need counseling based on their rate of traffic accidents and citizens' complaints; locate where prehistoric toolmakers found their stones; analyze writing styles; attribute the 1613 play *The Two Noble Kinsmen* to Shakespeare and his colleague John Fletcher; and adjust telescope mirrors to eliminate the atmospheric turbulence that makes stars appear to twinkle.

▶**358** *Wit writ in stone.* Encoded in this Latin plaque on Ernest Ludwig's house in Darmstadt, Germany, is the date when it was erected: "Ernest Ludwig, Count of Hesse, erected this palace where other buildings had been destroyed by fire." But where's the date?

The capitalized Roman numerals add up to the year 1715. Embedding construction dates in Latin prose was a popular eighteenth-century game—popular among counts who could build palaces, that is. An extremely witty bunch, eh?

> AB ERNESTO LVDoVICo
> LaNDGRaVIo HASSIÆ
> PRÆSENS ARX
> LoCo ALTERIVs
> VVLCaNI FVRORE ABREPTÆ
> EXstrVCta EST

►359 **Mathematically speaking, a knot is a closed curve without loose ends. It twists and turns through space like a tangle of yarn with its ends joined. The difference is that mathematical yarn has no thickness, stiffness, tension, or friction. Setting aside your Boy Scout and Girl Scout handbooks, guess how many different knots have been identified.**

Far more than 13,000. You can make that many different knots without having to cross one strand over another more than 13 times, without tying macrame-style one knot after another in the same string, and without even distinguishing knots from their mirror images. Only recently have mathematicians learned a way to estimate or calculate how difficult untying a particular knot will be. The technique should prove useful to physicists and biologists.

►360 **When is an enzyme like Alexander the Great?**

When it's a topoisomerase enzyme that cheats and unknots DNA by cutting it apart. When Alexander the Great learned that whoever could undo the Gordian knot would rule Asia, he crudely sliced open the knot with his sword. Topoisomerases take a similar, slash-and-be-damned approach to untangling DNA. Unlike mathematical knots, DNA is not a closed curve; but like the Gordian knot, it can't be untied by pulling out its loose ends. There just isn't room enough inside the nucleus of the cell. The only way to untie it is to cheat: that is, to cut the strand. The enzyme not only slashes the DNA but also reconnects it so that a section that had passed over another now goes under. Needless to say, Alexander did not bother to retie the Gordian knot.

▶361 Applied mathematicians greatly improved the operation of water clocks and sundials in Alexander the Great's empire between 336 B.C. and 323 B.C. How were the updated timepieces used?

To limit lawyers' speechifying in court. Sound familiar?

▶362 What is the longest mathematical proof ever produced without a computer?

A 15,000-page proof, totaling about 500 articles, written by 100 mathematicians during the early 1980s. It classified finite, simple groups. Cynics said that only its general contractor understood the whole thing and that he had died. Increasingly, mathematicians use computers and experiments to test their work. Traditionalists rely on proofs, offering only logical steps to move from self-evident truths to irrefutable conclusions.

▶363 From clean rooms to clean compiling and clean-and-certifying, the computer industry talks a lot about keeping squeaky clean and clear. So why is it messing around with bacteria?

To make smaller (and hence faster) switches. Bacteriorhodopsin, a light-harvesting protein in the purple membrane of salt-loving bacteria, thrives in marshwater six times saltier than seawater. The protein is so plentiful that it can color marsh water purple.

When oxygen levels drop too low for normal respiration, the bacterium simply switches to its backup system: It grows the purple membrane and, harvesting sunlight, pumps a photon of light energy back and forth across the membrane. In this way it generates enough alternate energy to continue metabolizing food.

Bacteriorhodopsin *Halobacterium salinarium* has three main charms for computer designers: (1) Absorbing and emitting photons, it can signal in less than one picosecond, which is 10^{-12} second. (2) It can switch on and off 10 million times before collapsing of exhaustion. (3) It survives heat.

▶364 The "Turtles of Aegina" were coins issued around 400 B.C. by the Greek city of Aegina near Athens. The city's coat of arms appeared on the coin face. On the back of the coin was a mathematical formula. What was it?

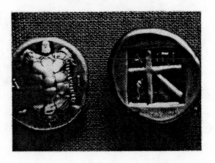

The binomial theorem: $(a+b)^2 = a^2 + b^2 + 2ab$.

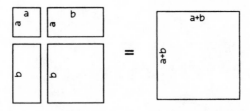

▶365 How can nations divide territory without warring over the spoils? How can divorcing couples divide assets amicably? How could governments allocate media resources without engendering further hostility?

By cutting cake like mathematicians, with the "envy-free cake division protocol." It's a 1995 variant on the child's "You cut; I choose" approach. *Warning:* Practically speaking, you may need a very large cake.

Step one: For four players, have the first player A cut the cake into *five* equal pieces, that is, pieces that she thinks are equal. With five players, cut the cake into nine pieces; for six players into 17, and so on.

Step two: Player B agrees that three or more pieces are tied for largest or trims one or two pieces to make what he thinks will be a three(or more)-way tie for largest.

Step three: Player C agrees that two or more pieces are already tied for largest, or she trims one piece to create what she regards as a two(or more)-way tie for largest.

Step four: The players choose pieces in the following order: D, C, B, and A. Those who trimmed a piece must take it, if it's available.

Step five: Reassemble trimmings and leftovers and repeat over and over again until only crumbs remain.

A mathematically pure process would tidy up the last crumbs (so that the entire cake is allocated in finitely many steps), but that involves 20 complex steps.

Lest this seem too arcane, remember how the four Allies split up Germany in 1945: They set aside Berlin as an extra fifth piece with trimmings to be divided later.

Recommended Reading

The numbers correspond to the numbers of the questions in the book.

Engineering and Technology

1 Fred Pearce. "Fogs Yield Drinking Water in the Desert." *New Scientist* 140 (Oct. 16, 1993): 19.

2 Robert Braham. "The Digital Backlot." *IEEE Spectrum* 32 (July 1995): 50–63; "Do Computers Dream of Cartoon Toys?" *New Scientist* 148 (Dec. 23/30, 1995): 11.

3 Tim Beardsley. "Testing's Toll." *Scientific American* 273 (Aug. 1995): 28.

4 Martin I. Jacobs. "Unzipping Velcro." *Scientific American* 274 (April 1996): 116.

5 Private communication, John L. Lumley, Cornell University.

6 John DeCicco and Marc Ross. "Improving Automotive Efficiency." *Scientific American* 271 (Dec. 1994): 52–57.

7 Dan Thisdell. "The Quick Route to an Emergency Stop." *New Scientist* 144 (Nov. 5, 1994): 20.

8 Edwin F. Meyer III. "Multiple-Car Pileups and the Two-Second Rule." *The Physics Teacher* 32 (Nov. 1994): 496–497.

9 Jon G. McGowan. "Tilting Toward Windmills." *Technology Review* 96 (July 1993): 40–46; Robert W. Thresher and Susan M. Hock. "Wind Systems for Electrical Power Production." *Mechanical Engineering* 116 (Aug. 1994): 68–72.

10 Joe P. Mahoney, University of Washington. Draft, *Washington State Department of Transportation Pavement Guide* vol. 1, (Jan. 1993).

11 Adrian Cottrill. "Auger Unveiled." *Offshore Engineer* (April 1994): 28–41; C. R. Enze et al. "Auger TLP Design, Fabrication, and Installation Overview." *Offshore Technology Conference Proceedings*, OTC 7615 (May 1994): 379–387; Rita Robison. "Bullwinkle's Big Brother." *Civil Engineering* 65 (July 1995): 44–47.

12 Toshiaki Ishii. "Elevators for Skyscrapers." *IEEE Spectrum* 31 (Sept. 1994): 42–46.

13 Jane Stevens. "Building with Bamboo." *Technology Review* 97 (Aug./Sept. 1994): 17–19; Marcelo Villegas. *Tropical Bamboo*. New York: Rizzoli, 1990.

14 Benjamin Miller. "Vital Signs of Identity." *IEEE Spectrum* 31 (Feb. 1994): 22–30.

15 See 14.

16 B. Bower. "Site Surrenders Fabric of Prehistoric Life." *Science News* 144 (July 24, 1993): 54; Brenda Fowler. "Find Suggests Weaving Preceded Settled Life." *New York Times*, May 9, 1995, B7; Jeff Hecht. "Strands from the Dawn of Time." *New Scientist* 139 (July 24, 1993): 8.

17 John Noble Wilford. "Human Ancestors' Tools Are Found in Africa." *New York Times*, April 25, 1995, B7.

18 Associated Press. "Oldest Tools Are Traced to Africa." *New York Times*, April 28, 1995, A13; Alison S. Brooks et al. "Dating and Context of Three Middle Stone Age Sites with Bone Points in the Upper Semliki Valley, Zaire." *Science* 268 (April 28, 1995): 548–553.

19 Gary Chamberlain. "Plastics Take Sports to New Peaks." *Design News* 50 (Aug. 28, 1995): 80–84.

20 Richard S. Peigler. "Wild Silks of the World." *American Entomologist* 39 (Fall 1993): 151–161.

21 Anne Simon Moffat. "Microbial Mining Boosts the Environment, Bottom Line." *Science* 264 (May 6, 1994): 778–779.

22 Robert R. Birge. "Protein-Based Three-Dimensional Memory." *American Scientist* 82 (July/Aug. 1994): 348–355; Robert R. Birge. "Protein-Based Computers." *Scientific American* 272 (March 1995): 90–95.

23 Kaigham J. Gabriel. "Engineering Microscopic Machines." *Scientific American* 273 (Sept. 1995): 150–153; Robert Langreth. "Scoping for Data." *Popular Science* 247 (Aug. 1995): 34; Ivars Peterson. "Microtools for Scaling Nanomountains." *Science News* 146 (April 1, 1995): 207.

24 Kirt R. Williams, private communication, University of California, Berkeley.

25 Trevor I. Williams, ed. *A History of Technology*, vol. 7. Oxford: Clarendon Press, 1978; Stuart Thorne. *The History of Food Preservation*. New York: Barnes and Noble, 1986; T. P. Coultate. *Food: The Chemistry of Its Components*. London: Royal Society of Chemistry, 1989.

26 Henry Petroski. "The Boeing 777." *American Scientist* 83 (Nov./Dec. 1995): 519–522.

27 John R. Hale. "The Lost Technology of Ancient Greek Rowing." *Scientific American* 274 (May 1996): 82–85.

28 Fred Pearce. "The International Grid." *New Scientist* 147 (July 8, 1995): 38–42.

29 See 28.

30 Charles Singer et al. *A History of Technology*, vol. 4. Oxford: Clarendon Press, 1958.

31 R. L. Weber. *A Random Walk in Science*. London: The Institute of Physics, 1973, p. 68.

Medicine and Health

32 Carl E. Bartecchi, Thomas D. MacKenzie, and Robert W. Schrier. "The Global Tobacco Epidemic." *Scientific American* 272 (May 1995): 44–51; Phyllida Brown. "Tobacco Kills One Smoker in Two." *New Scientist* 142 (Oct. 15, 1994): 4; Tara Patel. "Eastern Europe Heads for Smoking Catastrophe." *New Scientist* 143 (Oct. 22, 1994): 12.

33 "Addicted to Nicotine? Consider Snuff." *Science News* 146 (July 9, 1994): 30.

34 Susan J. Blumenthal. "Further Predictions on Medical Progress." *Scientific American* 273 (Sept. 1995): 140–141.

35 Lawrence D. Recht, Robert A. Lew, and William J. Schwartz. "Baseball Teams Beaten by Jet Lag." *Nature* 377 (Oct. 19, 1995): 583.

36 Nancy J. Alexander. "Future Contraceptives." *Scientific American* 273 (Sept. 1995): 136–141.

37 Natalie Angier. "Genetic Mutations Tied to Father in Most Cases." *New York Times*, May 17, 1994, C12.

38 Mark Ridley. "Why Presidents Have More Sons." *New Scientist* 144 (Dec. 3, 1994): 28–32; John F. Martin. "Hormonal and Behavioral Determinants of the Secondary Sex Ratio." *Social Biology* 42 (Fall/Winter 1995): 226–238.

39 Mertice Clark, Peter Karpiuk, and Bennet Galef. "Hormonally Mediated Inheritance of Acquired Characteristics in Mongolian Gerbils." *Nature* 364 (Aug. 19, 1993): 172; Mark Ridley. "Why Presidents Have More Sons." *New Scientist* 144 (Dec. 3, 1994): 28–32.

40 Mark Ridley. "Why Presidents Have More Sons." *New Scientist* 144 (Dec. 3, 1994): 28–32; John F. Martin. "Hormonal and Be-

havioral Determinants of the Secondary Sex Ratio." *Social Biology* 42 (Fall/Winter 1995): 226–238.

41 Lee Cronk. "Parental Favoritism Toward Daughters." *American Scientist* 81 (May/June 1993): 272–279.

42 Gail Vines. "Why Males Live Fast and Die Young." *New Scientist* 140 (Dec. 11, 1993): 18.

43 K. A. Fackelmann. "Male Rats Find Alcohol a Fertility Downer." *Science News* 146 (June 2, 1994): 6.

44 Harry Moore. "Sperm You Can Count On." *New Scientist* 124 (June 10, 1989): 38–41.

45 See 44.

46 N. M. Jacoby. "Krook's Dyslexia." *The Lancet* (Dec. 19/26, 1992): 1521–1522.

47–48 D. Christensen. "Pre-Columbian Mummy Lays TB Debate to Rest." *Science News* 145 (March 19, 1994): 181; Frank Ryan. *The Forgotten Plague: How the Battle Against Tuberculosis Was Won— and Lost.* Boston: Little, Brown, 1992, pp. 5–8; Wilmar L. Salo et al. "Identification of *Mycobacterium tuberculosis* DNA in a Pre- Columbian Peruvian Mummy." *Proceedings of the National Academy of Sciences* USA 91 (March 1994): 2091–2094; John Noble Wilford. "Tuberculosis Found to Be Old Disease in New World." *New York Times,* March 15, 1994, C1ff.

49 Herrman L. Blumgart. "Caring for the Patient." *New England Journal of Medicine* 270 (Feb. 27, 1964): 449–456.

50 Eliot A. Brenowitz. "Altered Perception of Species-Specific Song by Female Birds After Lesions of a Forebrain Nucleus." *Science* 251 (Jan. 18, 1991): 303–305.

51 Erwin H. Ackerknecht. *A Short History of Medicine.* Baltimore: Johns Hopkins Press, 1982.

52 *Statistical Abstract of the United States* 1992. U.S. Department of Commerce, Bureau of the Census.

53 Frank Ryan. *The Forgotten Plague: How the Battle Against Tuberculosis Was Won—and Lost*. Boston: Little, Brown, 1992, pp. 5–8.

54 Paul W. Ewald. "The Evolution of Virulence." *Scientific American* 268 (April 1993): 86–93; Paul E. Ewald. *Evolution of Infectious Disease*. Oxford: Oxford University Press, 1994.

55 See 54.

56 See 54.

57 Lyall Watson. "The Birdman of Batavia." *The Sciences* 29 (Jan./Feb. 1989): 36–37.

58 Robin Marantz Henig. "Flu Pandemic." *New York Times Magazine*, Nov. 29, 1992, 28ff; Robin Marantz Henig. *A Dancing Matrix: Voyages Along the Viral Frontier*. New York: Alfred A. Knopf, 1992.

59 W. French Anderson. "Gene Therapy." *Scientific American* 273 (Sept. 1995): 124–128.

60 Susan J. Blumenthal. "Further Predictions on Medical Progress." *Scientific American* 273 (Sept. 1995): 140–141.

61 Warren E. Leary. "Tissue Near Base of Skull Is Tied to Some Headaches." *New York Times*, Feb. 19, 1995, A12.

62 Sandra Blakeslee. "Old Accident Points to Brain's Moral Center." *New York Times*, March 24, 1994, B5; Hanna Damasio et al. "The Return of Phineas Gage: Clues About the Brain from the Skull of a Famous Patient." *Science* 264 (May 20, 1994): 1102–1105.

63 Fred Pearce. "The Rise and Rise of Pakistan's People." *New Scientist* 150 (April 27, 1996): 7; Population Action International, 1120 19th St. NW, Suite 550, Washington, DC 20036.

64 Bruce Bower. "Here Comes the Sun." *Science News* 142 (July 25, 1992): 62; David H. Avery et al. "Morning or Evening Bright Light Treatment of Winter Depression? The Significance of Hypersomnia." *Biological Psychiatry* 29 (1991):117–126; Richard J. Wurtman and Judith J. Wurtman. "Carbohydrates and Depression." *Scientific American* 260 (Jan. 1989): 68–75.

65 H. A. Waldron. *Medical History* 17 (1973): 391. Cited by Richard F. Mould, *Mould's Medical Anecdotes*. Bristol, England: Adam Hilger Ltd., 1984.

66 Gerald Weissmann. *The Woods Hole Cantata*. New York: Dodd Mead, 1985.

67 R. Weiss. "Women's Skills Linked to Estrogen Levels." *Science News* 134 (Nov. 26, 1988): 341.

68 Barry R. Bloom and Christopher J. L. Murray. "Tuberculosis: Commentary on a Reemergent Killer." *Science* 257 (Aug. 21, 1992): 1055–1063.

69 John Ding, E. Young, and Zanvil A. Cohn. "How Killer Cells Kill." *Scientific American* 258 (Jan. 1988): 38–44.

Ecology and Animal Behavior

70 E. Paul Catts. "Sex and the Bachelor Bot." *American Entomologist* 40 (Fall 1994): 153–160.

71 Curtis N. Runnels. "Environmental Degradation in Ancient Greece." *Scientific American* 272 (March 1995): 96–99.

72 J. Molloy and A. Davis. *Setting Priorities for the Conservation of New Zealand's Threatened Plants and Animals*. Wellington, New Zealand: Department of Conservation, 1994; Noel D. Vietmeyer. "The Salvation Islands." *Science and the Future 1993*. Chicago: Encyclopaedia Britannica, Inc., 1992.

73 Noel D. Vietmeyer. "The Salvation Islands." *Science and the Future 1993*. Chicago: Encyclopaedia Britannica, Inc., 1992; Clare Putnam. "Life Returns to Island After Rat Massacre." *New Scientist* 140 (Oct. 23, 1993): 15.

74 Michael Steele and Peter Smallwood. "What Are Squirrels Hiding?" *Natural History* 103 (Oct. 1994): 40–45.

75 Richard Monastersky. "A Question of Crushers." *Science News* 144 (Dec. 11, 1993): 396–397.

76 Nicholas B. Davies. "Backyard Battle of the Sexes." *Natural History* 104 (April 1995): 68–73; Nicholas B. Davies. *Dunnock Behaviour and Social Evolution.* New York: Oxford University Press, 1992.

77 See 76.

78 John Seidensticker. "Mountain Lions Don't Stalk People." *Smithsonian* 22 (Nov. 1992): 113–122; William K. Stevens. "Survival of the Big Cats Brings Conflict with Man." *New York Times,* Aug. 2, 1994, C1, C4.

79 See 78.

80 Howard B. Quigley. "Encounters with a Silent Predator." *Natural History* 103 (Dec. 1994): 57; Ian Ross. "Lions in Winter." *Natural History* 103 (Dec. 1994): 52–57; John Seidensticker. "Mountain Lions Don't Stalk People." *Smithsonian* 22 (Nov. 1992): 113–122; William K. Stevens. "Survival of the Big Cats Brings Conflict with Man." *New York Times,* Aug. 2, 1994, C1, C4.

81 T. Adler. "Deep-Sea Sponge Reaches Out, Devours." *Science News* 147 (Feb. 4, 1995): 69; Michelle Kelly-Borges. "Sponges Out of Their Depth." *Nature* 373 (Jan. 26, 1995): 284; J. Vacelet and N. Boury-Esnault. "Carnivorous Sponges." *Nature* 373 (Jan. 26, 1995): 333–335.

82 Tim Clutton-Brock. "Counting Sheep." *Natural History* 103 (March 1994): 29–34; A. Hansson-Lennart. "The Lemming Phenomenon." *Natural History* 98 (Dec. 1989): 38.

83 Malcolm W. Browne. "Eggs on Feet and Far from Shelter, Male Penguins Do a Shuffle." *New York Times,* Sept. 27, 1994, B5, B9; Graham Robertson. "Chaotic Cuddlers." *Natural History* 103 (Sept. 1994): 78–79.

84 Natalie Angier. "A Farming Ant and Its Fungus Are Ancient Cohabitants." *New York Times,* Dec. 13, 1994, B5; Ignacio H. Chapela et al. "Evolutionary History of the Symbiosis Between Fungus-Growing Ants and Their Fungi." *Science* 266 (Dec. 9, 1994): 1691–1694; B. Hölldobler and E. O. Wilson. *The Ants.* Cambridge, Mass.: Harvard University Press/Belknap Press, 1990.

85 Charles C. Mann. "Fire Ants Parlay Their Queens into a Threat to Biodiversity." *Science* 263 (March 18, 1994): 1560–1561.

86 David Concar. "Where There's Ants." *New Scientist* 144 (Oct. 29, 1994): 47.

87 Leigh Dayton. "The Land Where Snakes Are Top Dog." *New Scientist* 141 (March 12, 1994): 14.

88 Terry Dawson. "Kangaroos, the Kings of Cool." *Natural History* 104 (April 1995): 39–45.

89 David Concar. "Mystery of the Missing Mammals." *New Scientist* 144 (Oct. 29, 1994): 44–47.

90 Clare Putnam. "Fatal Attraction of Fungus the Parasite." *New Scientist* 142 (June 11, 1994): 20; Anders Møller. "A Fungus Infecting Domestic Flies Manipulates Sexual Behaviour of Its Host." *Behavioural Ecology and Sociobiology* 33, no. 6 (1993): 403–407.

91 Stuart Pimm. "What the Woods Won't Whisper." *The Sciences* 47 (May/June 1994): 16–19.

92 John Noble Wilford. "First Settler Domesticated Pigs Before Crops." *New York Times,* May 31, 1994, C1.

93 Tim Birkhead. "How Collectors Killed the Great Auk." *New Scientist* 142 (May 28, 1994): 24–27; Tim Birkhead. *Great Auk Islands.* London: Poyser, 1993; Bill Montevecchi. "The Great Auk Cemetery." *Natural History* 103 (Aug. 1994): 6–8; Bill Montevecchi. *Newfoundland Birds.* Cambridge, England: Nuttall Press, 1987.

94 Adrian Barnett. "Africa's Wild Dogs Pussyfoot Round the Big Cats." *New Scientist* 142 (April 30, 1994): 17; Joshua Ginsberg and Mary Cole. "Wild at Heart." *New Scientist* 144 (Nov. 19, 1994): 34–39.

95 David C. Houston. "To the Vultures Belong the Spoils." *Natural History* 103 (Sept. 1994): 35–41.

96 Debora MacKenzie. "Saving Haiti's Bacon." *New Scientist* 139 (July 17, 1993): 35–38.

97 William H. Karasov. "In the Belly of the Bird." *Natural History* 102 (Nov. 1993): 32–37.

98 Clive G. Jones and Moshe Shachak. "Desert Snails' Daily Grind." *Natural History* 103 (Aug. 1994): 56–61; Clive G. Jones and Moshe Shachak. "Fertilization of the Desert Soil by Rock-Eating Snails." *Nature* 346 (Aug. 1990): 839–841.

99 Robert A. Browne. "Sex and the Single Brine Shrimp." *Natural History* 102 (May 1993): 35–39.

100 See 99.

101 Monty Priede. "The Sea Snakes Are Coming." *New Scientist* 128 (Nov. 10, 1990): 29–33.

102 Kathy A. Fackelmann. "Loafing at the Landfill." *Science News* 145 (April 16, 1994): 252–253; Jerrold L. Belant et al. "Importance of Landfills to Nesting Herring Gulls." *The Condor* 95 (Nov. 1993): 817–830.

103 Bernd Heinrich and John Marzluff. "Why Ravens Share." *American Scientist* 83 (July/Aug. 1995): 342–349; Bernd Heinrich. *Ravens in Winter.* New York: Summit Books, 1989.

104 Bernd Heinrich. "Artful Diners." *Natural History* 89 (June 1980): 42–51; Pamela G. Real et al. "Discrimination and Generalization of Leaf Damage by Blue Jays." *Animal Learning and Behavior* 12 (May 1984): 202–208.

105 John H. Cushman Jr. "Freshwater Mussels Facing Mass Extinction." *New York Times*, Oct. 3, 1995, C1.

106 Malcolm W. Browne. "Scientists Hope More New Species Will Be Discovered in Vietnam." *New York Times*, May 3, 1994, C4; Tim Hilchey. "New Kangaroo Species Is Reported." *New York Times*, July 26, 1994, C4; Les Line. "A Newfound Mammal of Philippine Treetops Gets High-Flown Name." *New York Times*, Feb. 20, 1996, B6.

107 Carol Kaesuk Yoon. "Woolly Flying Squirrel, Long Thought Extinct, Shows Up in Pakistan." *New York Times,* March 14, 1995, B8; Perla Magsalay et al. "Extinction and Conservation on Cebu." *Nature* 373 (Jan. 26, 1994): 294.

Zoology and Animal Physiology

108 Johnson Timson. "Nature's Portable Sperm Bank." *New Scientist* 142 (April 2, 1994): 14; Tim Birkhead and Anders Møller. *Sperm Competition in Birds.* Orlando, Fla.: Academic Press, 1992.

109 Natalie Angier. "Sex and the Fruit Fly: Price of Promiscuity Is Premature Death." *New York Times,* Jan. 24, 1995, B7; Tracey Chapman et al. "Cost of Mating in *Drosophila melanogaster* Females Is Mediated by Male Accessory Gland Products." *Nature* 373 (Jan. 19, 1995): 241–244; Laurent Keller. "All's Fair When Love Is War." *Nature* 373 (Jan. 19, 1995): 190–191.

110 T. Adler. "Secret to Birds' Mating Score: Speedy Sperm." *Science News* 148 (Oct. 7, 1995): 231; T. R. Birkhead et al. "Ejaculate Quality and the Success of Extra-Pair Copulations in the Zebra Finch." *Nature* 377 (Oct. 5, 1995): 422–423.

111 Jussi Viitala et al. "Attraction of Kestrels to Vole Scent Marks Visible in Ultraviolet Light." *Nature* 373 (Feb. 2, 1995): 425–427.

112 Natalie Angier. "Clue Found to a Puzzle About Single-Sex Fish." *New York Times,* Jan. 25, 1994, C4; I. Schlupp, C. A. Marler, and M. J. Ryan. "Benefit to Male Sailfin Mollies of Mating with Heterospecific Females." *Science* 263 (Jan. 21, 1994): 373–374.

113 Natalie Angier. "Canary Chicks: Not All Created Equal." *New York Times,* Jan. 25, 1994; C4, C8.

114 Vincent Kiernan. "Intimate Secrets of the Snake's Forked Tongue." *New Scientist* 141 (March 26, 1994): 8; Kurt Schwenk. "Why Snakes Have Forked Tongues." *Science* 263 (March 18,1994): 1573–1577; Kurt Schwenk. "The Serpent's Tongue." *Natural History* 104 (April 1995): 48–55.

115 Masakazu Konishi. "Listening with Two Ears." *Scientific American* 268 (April 1993): 66–73.

116 Gerald J. FitzGerald. "The Reproductive Behavior of the Stickleback." *Scientific American* 268 (April 1993): 80–85.

117 "Picky Eaters in Monterey Bay Who Dabble in Petty Theft." *New York Times,* March 29, 1994, B8.

118 See 117.

119 C. Sue Carter and Lowell L. Getz. "Monogamy and the Prairie Vole." *Scientific American* 268 (June 1993): 100–106; Kathy A. Fackelmann. "Hormone of Monogamy." *Science News* 144 (Nov. 27, 1993): 360–365. "Vole Love." *Science* 265 (Sept. 30, 1994): 2007.

120 Natalie Angier. "Lactating Male Bats Found in Malaysia, Researchers Report." *New York Times,* Feb. 23, 1994, A1; K. A. Fackelmann. "Real Males That Lactate: A Batty Story." *Science News* 145 (March 5, 1994): 148; Charles M. Francis et al. "Lactation in Male Fruit Bats." *Nature* 367 (Feb. 24, 1994): 691–692; Usha Lee McFarling. "Rare Fruit Bats Found to Practice 'Midwifery.'" *Boston Globe,* June 6, 1994.

121 Clare Putnam. "Bat Mothers Share the Birth Experience." *New Scientist* 142 (June 18, 1994): 18.

122 Jonathan Beard. "Ants That Are Quick on the Jaw." *New Scientist* 140 (Dec. 4, 1993): 14; W. Gronenberg, J. Tautz, and B. Hölldobler. "Fast Trap Jaws and Giant Neurons in the Ant *Odontomachus.*" *Science* 262 (Oct. 22, 1993): 561–563; "The Fastest Jaw in the West." *Science News* 144 (Nov. 6, 1993): 302.

123 Neville Peat. *The Incredible Kiwi.* Auckland, New Zealand: Random Century, 1990; Michael Taborsky and Barbara Taborsky. "The Kiwi's Parental Burden." *Natural History* 102 (Dec. 1993): 50–57, 94.

124 Jared M. Diamond. "Nine Hundred Kiwis and a Dog." *Nature* 338 (April 13, 1989): 544; Neville Peat. *The Incredible Kiwi.* Auckland, New Zealand: Random Century 1990; Michael

Taborsky and Barbara Taborsky. "The Kiwi's Parental Burden." *Natural History* 102 (Dec. 1993): 50–57, 94.

125 Richard A. Lutz and Janet R. Voight. "Close Encounter in the Deep." *Nature* 371 (Oct. 13, 1994): 563; Rosie Mestel. "Torrid Sex Scenes Puzzle Octopus Experts." *New Scientist* 143 (Oct. 22, 1994): 12.

126 Knut Schmidt-Nielsen. "How Are Control Systems Controlled?" *American Scientist* 82 (Jan./Feb. 1994): 38–44; C. Claiborne Ray. "Drinking Salt Water." *New York Times,* June 15, 1993, B6.

127 Brian C. Bowen and John C. Avise. "Tracking Turtles through Time." *Natural History* 103 (Dec. 1994): 36–42; Kenneth J. Lohmann. "How Sea Turtles Navigate." *Scientific American* 26 (Jan. 1992): 100–106; Kenneth J. Lohmann and Catherine M. Fittinghoff Lohmann. "Detection of Magnetic Field Intensity by Sea Turtles." *Nature* 380 (March 7, 1996): 59–61; E. Pennisi. "Light May Aid Birds' Magnetic Orientation." *Science News* 144 (Aug. 7, 1993): 86; Lisa Seachrist. "Sea Turtles Master Migration with Magnetic Memories." *Science* 264 (April 29, 1994): 661–662; Laura Spinney. "Honeybees Iron Out Their Directional Problems." *New Scientist* 143 (July 23, 1994): 16.

128 Larry G. Marshall. "The Terror Birds of South America." *Scientific American* 270 (Feb. 1994): 90–95.

129 Sarah Bunney. "Most Ancient Human Came from Afar." *New Scientist* 154 (Oct. 1, 1994): 16; Blake Edgar. "Digging Up the Family Bones." *BioScience* 45 (Nov. 1995): 659–662; Joshua Fischman. "Putting Our Oldest Ancestors in Their Proper Place." *Science* 265 (Sept. 30, 1994): 2011–2012; "Team Unearths Oldest Known Human Ancestor." *Science News* 146 (Oct. 1, 1994): 212–213; "The Age of Australopithecines." *Nature* 372 (Nov. 3, 1994): 31–32.

130 Craig B. Stanford. *The Colobus and the Chimpanzee.* Cambridge: Harvard University Press: in press; Craig B. Stanford. "To Catch a Colobus." *Natural History* 104 (Jan. 1995): 48–55; Craig B.

Stanford. "Chimpanzee Hunting Behavior and Human Evolution." *American Scientist* 83 (May/June 1995): 256–262.

131 Natalie Angier. "Cotton-Top Tamarins: Cooperative, Pacifist and Close to Extinction." *New York Times,* Sept. 13, 1994, B5.

132 Frans B. M. de Waal. "Bonobo Sex and Society." *Scientific American* 272 (March 1995): 82–88.

133 Everhard Home. *Lectures on Comparative Anatomy: in which are explained the preparations in the Hunterian collection,* vol. 6. London: W. Bulmer for G. & W. Nicol, 1828; Tierney Thys. "Swimming Heads." *Natural History* 103 (Aug. 1994): 36–38.

134 Natalie Angier. "Flyspeck on a Lobster Lip Turns Biology on Its Ear." *New York Times,* Dec. 14, 1995, A1; Simon Conway Morris. "A New Phylum from the Lobster's Lips." *Nature* 378 (Dec. 14, 1995): 661; Peter Funch and Reinhardt Møbjerg Kristensen. "Cycliophora Is a New Phylum with Affinities to Entoprocta and Ectoprocta." *Nature* 378 (Dec. 14, 1995): 711–714.

135 Fred Bruemmer. "Low, Lean Killing Machine." *Natural History* 102 (Jan. 1993): 54–61; David Tomlinson. "Penguins Are Not the Only Food." *New Scientist* 143 (June 11, 1994): 45; Euan Young. *Skua and Penguin.* Cambridge, England: Cambridge University Press, 1994.

136 Peter Aldhous. "Fish Nose Ahead of Humans in Greedy Brain Stakes." *New Scientist* 149 (March 9, 1996): 20.

137 Alan Walker and Mark Teaford. "The Hunt for Proconsul." *Scientific American* 260 (Jan. 1989): 76.

138 Steven N. Austad. "The Adaptable Opossum." *Scientific American* 258 (Feb. 1988): 98–104.

139–145 Robert T. Bakker. "The Dinosaur Renaissance." *Science and the Future 1993.* Chicago: Encyclopaedia Britannica, Inc., 1993; Blake Edgar. "The Case of the Cold-Loving Dinosaurs." *Pacific Discovery* 47 (Summer 1994): 3; David E. Fastovsky and David B. Weishampel. *The Evolution and Extinction of the Dinosaurs.* Cambridge, England: Cambridge University Press, 1996.

Plant Science

146 John Noble Wilford. "Long Before the Plants Began to Blossom, Insects Knew How to Use Them." *New York Times,* Aug. 3, 1993, B5; C. C. Labardeira and J. J. Sepkoski Jr. "Insect Diversity in the Fossil Record." *Science* 216 (July 23, 1993): 310–315.

147 Carl Franklin. "Living Fungi Found in Ötzi's Boots." *New Scientist* 141 (March 19, 1994): 10.

148 Kevin S. Richter et al. "Differential Abortion in the Yucca." *Nature* 372 (Aug. 17, 1994): 557.

149 Gina Kolata. "In Ancient Times, Flowers and Fennel for Family Planning." *New York Times,* March 8, 1994, C5; John Riddle. *Contraception and Abortion from the Ancient World to the Renaissance.* Cambridge: Harvard University Press, 1992.

150 Royce Rensberger. "Plant-Derived Birth Control Is Ancient History." *Seattle Times,* Aug. 30, 1994, A14.

151 James H. Tumlinson, W. Joe Lewis, and Louise E. M. Vet. "How Parasitic Wasps Find Their Hosts." *Scientific American* 266 (March 1993): 100–106.

152 Kathleen K. Treseder, Diane W. Davidson, and James R. Ehleringer. "Absorption of Ant-Provided Carbon Dioxide and Nitrogen by a Tropical Epiphyte." *Nature* 375 (May 11, 1995): 137–139.

153 Nicola Perrin. "A Never-Ending Feast." *Scientific American* 272 (Nov. 1995): 24.

154 Bernd Heinrich. "Of Bedouins, Beetles, and Blooms." *Natural History* 103 (May 1994): 52–59.

155 Elizabeth Pennisi. "Static Evolution." *Science News* 145 (March 12, 1994): 168–169; J. W. Schopf, ed. *The Oldest Fossils and What They Mean.* Boston: Jones & Bartlett, 1992, pp. 29–63.

156 Edward S. Ross. "What's a *Welwitschia?*" *Pacific Discovery* 47 (Fall 1994): 44–45.

157 Wayne P. Armstrong. "The Tiniest Titan." *Pacific Discovery* 42 (Summer 1989): 33–38.

158 William P. Jacobs. "Caulerpa." *Scientific American* 271 (Dec. 1994): 100–105.

159 Theodore H. Fleming. "Cardon and the Night Visitors." *Natural History* 103 (Oct. 1994): 58–64.

160 See 159.

161 Cecilia C. Mettler and Fred A. Mettler. *History of Medicine*. London: Blakiston, 1947.

162 Janet Raloff. "What's In a Cigarette?" *Science News* 145 (May 21, 1995): 330–331.

163 Samuel M. Scheiner and José M. Rey-Benayas. "Some Like It Hot—And Cold." *Scientific American* 271 (Nov. 1994): 30; Samuel M. Scheiner and José M. Rey-Benayas. "Global Patterns of Plant Diversity." *Evolutionary Ecology* 8 (Aug. 1994): 331–347.

164 Steven R. King. "First in the Ancient Americas." *Pacific Discovery* 45 (Winter 1992): 23–31.

165 See 164.

166 See 164.

167 Jared Diamond. "Spacious Skies and Tilted Axes." *Natural History* 103 (May 1994): 16–23.

168 See 167.

169 Donald M. Ball, Jeffrey F. Pedersen, and Garry D. Lacefield. "The Tall-Fescue Endophyte." *American Scientist* 81 (July/Aug. 1993): 370–379.

170 R. Monastersky. "Rooting Around for Missing Carbon." *Science News* 146 (Sept. 17, 1994): 180; D. Schimel et al. "CO_2 and the Carbon Cycle." In *Climate Change 1994*, ed. J. T. Houghton et al. Intergovernmental Panel on Climate Change. Cambridge, England: Cambridge University Press, 1995, pp. 35–71.

171 Robert Sommer. "Why I Will Continue to Eat Corn Smut." *Natural History* 104 (Jan. 1995): 18–22.

172 William K. Stevens. "Invading Weed Makes a Bid to Become a New Kudzu." *New York Times,* Aug. 16, 1994, C4.

173 "Rarest U. S. Tree Under Fungus Attack." *Science News* 149 (Jan. 6, 1996): 15; Sasha Nemecek. "Rescuing an Endangered Tree." *Scientific American* 274 (March 1996): 22.

174 Karen F. Schmidt. "Good Vibrat-ions." *Science News* 139 (Dec. 14, 1991): 392–394.

175 Richard Mack. "Catalog of Woes." *Natural History* 99 (March 1990): 51–52.

176 Carol Kaesuk Yoon. "Warming Moves Plants Up Peaks, Threatening Extinction." *New York Times,* June 21, 1994, C4; G. Grabherr, M. Gottfried, and H. Pauli. "Climate Effects on Mountain Plants." *Nature* 396 (June 9, 1994): 448.

177 B. Wuethrich. "Forests in the Clouds Face Stormy Future." *Science News* 144 (July 10, 1993): 23.

178 Edward O. Wilson. *The Diversity of Life.* Cambridge: Harvard University Press, 1992; R. S. Nadgauda, V. A. Parasharami, and A. F. Mascarenhas. "Flowering and Seeding Behaviour in Tissue-Cultured Bamboos." *Nature* 344 (March 22, 1990): 335–336.

179 Natalie Angier. "Plants Paths to Devouring Insects." *New York Times,* Sept. 15, 1992, B9.

180 Gary A. Strobel. "Biological Control of Weeds." *Scientific American* 265 (July 1991): 72ff.

181 John S. Hardman. "More Than Just a Dry Drupe." *Chemistry in Britain* 28 (Aug. 1992): 711–714.

182 Richard Stone. "Surprise! A Fungus Factory for Taxol?" *Science* 260 (April 19, 1993): 154–155; Andrea Stierle, Gary Strobel, and Donald Stierle. "Taxol and Taxane Production by *Taxomyces andreanae,* an Endophytic Fungus of Pacific Yew." *Science* 260 (April 9, 1993): 214–216.

Earth Sciences

183–185 R. B. Alley et al. "Abrupt Increase in Greenland Snow Accumulation at the End of the Younger Dryas Event." *Nature* 362 (April 8, 1993): 527–529; Wallace S. Broecker. "Chaotic Climate." *Scientific American* 273 (Nov. 1995): 62–68; Cuffey et al. "Large Arctic Temperature Change at the Wisconsin-Holocene Glacial Transition." *Science* 270 (Oct. 20, 1995): 455–458; Richard G. Fairbanks. "Flip-flop End to Last Ice Age." *Nature* 362 (April 8, 1993): 495.

186 J. Robert Toggweiler. "The Ocean's Overturning Circulation." *Physics Today* 47 (Nov. 1994): 45–50.

187 See 186.

188 "New Theory Tells How Earth Grows Skin." *Science News* 145 (June 18, 1994): 397; Enrico Bonatti. "The Earth's Mantle Below the Oceans." *Scientific American* 270 (March 1994): 44–51.

189 C. D. Farrar et al. "Forest-Killing Diffuse CO_2 Emission at Mammoth Mountain as a Sign of Magmatic Unrest." *Nature* 376 (Aug. 24, 1995): 675–677; Stanley N. Williams. "Dead Trees Tell Tales." *Nature* 376 (Aug. 24, 1995): 644.

190 William J. Broad. "Clues Emerge to Rich Lodes of Diamonds." *New York Times*, Feb. 15, 1994, B5.

191 Rebecca Renner. "The Hottest Rocks on Earth." *New Scientist* 139 (July 24, 1993): 23–27.

192 Michael Bevis et al. "Geodetic Observations of Very Rapid Convergence and Back-arc Extension at the Tonga Arc." *Nature* 374 (March 16, 1995): 249–251; Harry W. Green II. "Solving the Paradox of Deep Earthquakes." *Scientific American* 271 (Sept. 1994): 64–71; Heidi Houston. "Deep Quakes Shake Up Debate." *Nature* 372 (Dec. 22/29, 1994): 724; Ray Ladbury. "Bolivian Quake Gives a Rare Glimpse of Earth's Interior." *Physics Today* 47 (Oct. 1994): 17–19.

193 W. I. Rose et al. "Ice in the 1994 Rabaul Eruption Cloud: Implications for Volcano Hazard and Atmospheric Effects." *Nature* 375 (June 8, 1995): 477–479.

194 William J. Broad. "Heavy Volcanic Eras Were Caused by Plumes from the Earth's Core." *New York Times,* Aug. 22, 1995, B7; Millard F. Coffin and Olav Eldholm. "Large Igneous Provinces." *Scientific American* 269 (Oct. 1993): 42–49.

195 See 194.

196 Cliff Frohlich. "Shaken to the Core." *New Scientist* 148 (Dec. 9, 1995): 43–46; Cliff Frohlich. "Deep Earthquakes." *Scientific American* 260 (Jan. 1989): 48–55.

197 Carl Franklin. "'Black Smokers Multiply on Ocean Floor." *New Scientist* 143 (Oct. 22, 1994): 20.

198 Robert Monastersky. "Light at the Bottom of the Ocean." *Science News* 145 (Jan. 1, 1994): 14.

199 Joe Cann and Cherry Walker. "Breaking New Ground on the Ocean Floor." *New Scientist* 140 (Oct. 30, 1993): 24–29; Christopher R. German and Martin V. Angel. "Watery Wastes." *Chemistry in Britain* 30 (July 1994): 560–561; Richard A. Lutz et al. "Rapid Growth at Deep-Sea Vents." *Nature* 371 (Oct. 20, 1994): 663–664.

200 T. Adler. "Bacteria Found Deep Below Ocean Floor." *Science News* 146 (Oct. 1, 1994): 215; R. John Parkes, "Deep Bacterial Biosphere in Pacific Ocean Sediments. *Nature* 371 (Sept. 29, 1994): 410–413.

201 "When Life First Sprouted on Land." *Science News* 145 (March 12, 1994): 173; Robert J. Horodyski and L. Paul Knauth. "Life on Land in the Precambrian." *Science* 265 (Jan. 28, 1994): 494–499.

202 "Megabergs Left Scars in Arctic." *Science News* 146 (Aug. 20, 1994): 127; Peter R. Vogt, Kathleen Crane, and Eirik Sundvor. "Deep Pleistocene Iceberg Plowmarks on the Yermak Plateau."

Geology 22 (May, 1994): 403–406; "Arctic Ice Sheet?" *Science* 265 (Aug. 5, 1994): 735.

203 Richard Monastersky. "'Great Green Wall' Dampens Gobi Dust Storms." *Science News* 145 (June 25, 1994): 406.

204 David Schoonmaker. "What Makes Permafrost Permanent?" *American Scientist* 81 (Nov./Dec. 1993): 527–528.

205 Marcia Barinaga. "Archaea and Eukaryotes Grow Closer." *Science* 264 (May 27, 1994): 1251; Robert M. Kelly, John A. Baross, and Michael W. W. Adams. "Life in Boiling Water." *Chemistry in Britain* 30 (July 1994): 555–558; Edward F. DeLong, Ke Ying Wu, Barbara B. Prézelin, and Raffael V. M. Jovine. "High Abundance of Archaea in Antarctic Marine Picoplankton." *Nature* 371 (Oct. 20, 1994): 695–697; Rosie Mestel. "Teeming Life in Ocean Deeps." *New Scientist* 142 (June 11, 1994): 19; Gary J. Olsen. "Archaea, Archaea, Everywhere." *Nature* 371 (Oct. 20, 1994): 657–658.

206 David Thomas and Gerhard Dieckmann. "Life in a Frozen Lattice." *New Scientist* 142 (June 11, 1994): 33–37.

207 Scott Faber. "Sky Rivers." *Discover* 15 (Jan. 1994): 34–35; Yong Zhu and Reginald E. Newell. "Atmospheric Rivers and Bombs." *Geophysical Research Letters* 21 (Sept. 1, 1994): 1999–2002.

208 William J. Broad. "Newfound Elves (Not Santa's) Blaze at the Very Edge of Space." *The New York Times,* Dec. 12, 1995, B5; Richard Monastersky. "Glowing Doughnuts Flash High Above Storms." *Science News* 148 (Dec. 23/30, 1995): 421.

209 "A Shocking Side to the Blizzard of '93." *Science News* 144 (Aug. 7, 1993): 95; Mark Malone, Global Atmospherics, Inc., private communication.

210 James A. Yoder et al. "A Line in the Sea." *Nature* 371 (Oct. 20, 1994): 689–691; T. Adler. "Microorganisms Create a Line in the Ocean." *Science News* 146 (Oct. 22, 1994): 263.

211 Tim Thwaites. "Are the Antipodes in Hot Water?" *New Scientist* 144 (Nov. 12, 1994): 21.

212 C. Hans Nelson and Kirk R. Johnson. "Whales and Walruses as Tillers of the Sea Floor." *Scientific American* 256 (Feb. 1987): 112–117.

213 Richard Monastersky. "Sahara Dust Blows over United States." *Science News* 148 (Dec. 23/30, 1995): 431.

214 David A. Greenberg. "Modeling Tidal Power." *Scientific American* 257 (Nov. 1987): 128–131; Roger Henri Charlier. *Tidal Energy.* New York: Van Nostrand, 1982, p. 151; Trevor I. Williams, ed. *A History of Technology,* vol. 6. Oxford, England: Clarendon Press, 1978, 217–218.

215 John S. Rinehart. *A Guide to Geyser Gazing.* Santa Fe: Hyper Dynamics, 1976.

216 Fred Pearce, "Ancient Lessons from Arid Lands." *New Scientist* 132 (Dec. 7, 1991): 42ff; William M. Denevan, "Hydraulic Agriculture in the American Tropics: Forms, Measures, and Recent Research." In *Maya Subsistence,* ed. Kent Flannery. New York: Academic Press, 1982, pp. 181–220.

217 Walter Sullivan, "Sea Floor Holds Story of Hawaiian Isles' Doom." *New York Times,* Dec. 3, 1991, C1.

218 Richard Monastersky. "Fire Beneath the Ice." *Science News* 143 (Feb. 13, 1993): 104–107.

219 Amos Nur. "And the Walls Came Tumbling Down." *New Scientist* 131 (July 6, 1991): 45–48.

220 Robert S. White. "Ancient Floods of Fire." *Natural History* 100 (April 1991): 51–60.

221 Daniel Hillel. "Lash of the Dragon." *Natural History* 100 (Aug. 1991): 31–36.

The Molecules of Life

222 Scott Pitnick, Greg S. Spicer, and Therese A. Markow. "How Long Is a Giant Sperm?" *Nature* 375 (May 11, 1995): 109; Tim-

othy L. Karr and Scott Pitnick. "The Ins and Outs of Fertilization." *Nature* 379 (Feb. 1, 1996): 405.

223 D. Joly, C. Bressac, and D. Lachaise. "Disentangling Giant Sperm." *Nature* 377 (Sept. 21, 1995): 202.

224 H. Hennakao Komlyama et al. "Transplanting a Unique Allosteric Effect from Crocodile into Human Haemoglobin." *Nature* 373 (Jan. 19, 1995): 244–246.

225 Svante Pääbo. "Ancient DNA." *Scientific American* 269 (Nov. 1993): 86, 92.

226 John Bonner. "It's a Dog's Life." *New Scientist* 144 (Nov. 5, 1994): 34–36.

227 R. Monastersky. "Birds: Lightweights in the Genetic Sense." *Science News* 148 (Oct. 7, 1995): 229; Austin L. Hughes and Marianne K. Hughes. "Small Genomes for Better Flyers." *Nature* 377 (Oct. 5, 1995): 391.

228 E. Pennisi. "Mice, Flies Share Memory Molecule." *Science News* 146 (Oct. 15, 1994): 244.

229 "Do Songbirds Sing of Alzheimer's?" *Science News* 148 (Aug. 26, 1995): 139.

230 Virginia Morell. "Pulling Hair from the Ground." *Science* 265 (Aug. 5, 1994): 741; Virginia Morell. "Decoding Chimp Genes and Lives." *Science* 265 (Aug. 26, 1994): 1172–1173.

231 H. D. Bradshaw et al. "Genetic Mapping of Floral Traits Associated with Reproductive Isolation in Monkeyflowers *(Mimulus)*." *Nature* 376 (Aug. 31, 1995): 762–764; Jerry A. Coyne. "Speciation in Monkeyflowers." *Nature* 376 (Aug. 31, 1995): 726–727.

232 Stephen Day. "Gene Decides What Tickles the Tastebuds." *New Scientist* 139 (Aug. 14, 1993): 14; Martha Guarna and Richard Borowsky. "Genetically Controlled Food Preference." *Proceedings of the National Academy of Sciences* 90 (June 1, 1993): 5257–5262.

233 Thomas P. Stossel. "The Machinery of Cell Crawling." *Scientific American* 271 (Sept. 1994): 54–63.

234 R. Monastersky. "Ancient Bacteria Brought Back to Life." *Science News* 147 (May 20, 1995): 308.

235 Michelle Hoffman. "Some Old Genes Just Won't Quit." *American Scientist* 83 (May/June 1995): 235–236.

236 K. S. J. "How the Mouse's Tale Began Life in India." *Nature* 377 (Sept. 21, 1995): 188.

237 Virginia Morell. "Decoding Chimp Genes and Lives." *Science* 265 (Aug. 26, 1994): 1172–1173.

238 Russell F. Doolittle and Peer Bork. "Evolutionarily Mobile Modules in Proteins." *Scientific American* 269 (Oct. 1993): 50–56.

239 Nicholas Wade. "First Sequencing of Cell's DNA Defines Basis of Life." *New York Times,* Aug. 1, 1995, B5ff.

240 Bob Holmes. "Inside the Gene Machine." *New Scientist* 150 (April 27, 1996): 26–29.

241 Andy Coghlan. "Gene Dream Fades Away." *New Scientist* 148 (Nov. 25, 1995): 14–15.

242 William W. Hauswirth. "Dead Men's Molecules." *Science and the Future 1994.* Chicago: Encyclopaedia Britannica Yearbook, 1994; Svante Pääbo. "Ancient DNA." *Scientific American* 269 (Nov. 1993): 86–92.

243 John Noble Wilford. "First Branch in Life's Tree Was 2 Billion Years Ago." *New York Times,* Jan. 30, 1996, B5.

244 David A. Grimaldi. "Captured in Amber." *Scientific American* 274 (April 1996): 84–91; Philip J. Hilts. "Expedition to Far New Jersey Finds Trove of Amber Fossils." *New York Times* Jan. 20, 1996, B10.

245 See 244.

246 Akishinonamiya Fumihito et al. "One Subspecies of the Red Junglefowl (*Gallus gallus gallus*) Suffices as the Matriarchic An-

cestor of All Domestic Breeds." *Proceedings of the National Academy of Sciences USA* 91 (Dec. 1994): 12505–12509.

247 Karen Schmidt. "A New Ant on the Block." *New Scientist* 148 (Nov. 4, 1995): 28–31.

248 Bob Holmes. "Message in a Genome?" *New Scientist* 147 (Aug. 12, 1995): 30–33; Mark Pagel and Rufus A. Johnstone. "Variation Across Species in the Size of the Nuclear Genome Supports the Junk-DNA Explanation for the C-Value Paradox." *Proceedings of the Royal Society London,* Series B, 249 (1992): 119–124.

249 William Menasco and Lee Rudolph. "How Hard Is It to Untie a Knot?" *American Scientist* 83 (Jan./Feb. 1995): 38–49.

250 Sarah Bunney. "DNA Survey Sets Basques Apart." *New Scientist* 141 (March 19, 1994): 17; Francesc Calafell and Jaume Bertranpetit. "Principal Component Analysis of Gene Frequencies and the Origin of Basques." *American Journal of Physical Anthropology* 93 (Feb. 1994): 201–215.

251 Michael J. Novacek. "Where Do Rabbits and Kin Fit In?" *Nature* 379 (Jan. 25, 1996): 299–300; Dan Grauer, Laurent Duret, and Manolo Gouy. "Phylogenetic Position of the Order Lagomorpha (Rabbits, Hares and Allies)." *Nature* 379 (Jan. 25, 1996): 333–335.

Chemistry

252 Tara Patel. "Camel's Cheese Gives Nourishment to Nomads." *New Scientist* 139 (July 31, 1993): 8.

253 Luigi Garlaschelli, Franco Ramaccini, and Sergio Della Sala. "A 'Miracle' Diagnosis." *Chemistry in Britain* 30 (Feb. 1994): 123–125; Luigi Garlaschelli, Franco Ramaccini, and Sergio Della Sala. "Working Bloody Miracles." *Nature* 353 (Oct. 10, 1991): 507.

254 Cherry A. Murray and David G. Grier. "Colloidal Crystals." *American Scientist* 83 (May/June 1995): 238–245; Private communication, John Berg, University of Washington.

255 Nigel P. Freestone, Paul S. Phillips, and Ray Hall. "Having the Last Gas." *Chemistry in Britain* 30 (Jan. 1994): 48–50.

256 Debora MacKenzie. "Chewing Gum Gave Stone Age Punk a Buzz." *New Scientist* 139 (Sept. 18, 1993): 7.

257 Michael Boppré. "Sex, Drugs, and Butterflies." *Natural History* 103 (Jan. 1994): 26–33; Clare Putnam. "Wizards of Chemistry on the Wing." *New Scientist* 141 (Feb. 19, 1994): 16; Stefan Schulz, Michael Boppré, and R. Vane-Wright. "Specific Mixtures of Secretions from Male Scent Organs of African Milkweed Butterflies." *Philosophical Transactions of the Royal Society of London,* Series B, 342 (Oct. 29, 1993): 161–181.

258 Stephanie Pain. "Pests Leave Lasting Impression on Plants." *New Scientist* 145 (March 4, 1995): 13.

259 John Emsley. "Ancient World Was Poisoned by Lead." *New Scientist* 143 (Oct. 1, 1994): 14; Sungmin Hong et al. "Greenland Ice Evidence of Hemispheric Lead Pollution 2 Millennia Ago by Greek and Roman Civilizations." *Science* 265 (Sept. 28, 1994): 1841.

260 Henry Gee. "Neighborhood Watch." *Nature* 377 (Sept. 28, 1995): 289; Masato Ono et al. "Unusual Thermal Defence by a Honeybee Against Mass Attack by Hornets." *Nature* 377 (Sept. 28, 1995): 334–336.

262 Seth J. Putterman. "Sonoluminescence: Sound into Light." *Scientific American* 272 (Feb. 1995): 46–51; Kenneth J. Suslick. "The Chemistry of Ultrasound." *Science and the Future 1994.* Chicago: Encyclopaedia Britannica, Inc., 1994, pp. 139–155.

263 Carol A. Phillips, Toni Gladding, and Susan Maloney. "Clouds with a Quicksilver Lining." *Chemistry in Britain* 30 (Aug. 1994): 646–656.

264 Scott R. Smedley and Thomas Eisner. "Sodium Uptake by Puddling in a Moth." *Science* 270 (Dec. 15, 1995): 1816–1819; Kai Wu. "Pass the Salt, Please." *Scientific American* 274 (March 1996): 30.

265 Rosie Mestel. "How Blue Genes Could Green the Cotton Industry." *New Scientist* 139 (July 31, 1993): 7; "100% Natural Polyester?" *Science* 266 (Dec. 16, 1994): 1811.

266 Rachel Nowak. "Nicotine Scrutinized as FDA Seeks to Regulate Cigarettes." *Science* 263 (March 18, 1994): 1555–1556.

267 "Hot Answers to Some 'Bad Hair' Problems." *Science News* 144 (Dec. 11, 1993): 391; Susan P. Detwiler et al. "Bubble Hair." *Journal of the American Academy of Dermatology* 30 (Jan. 1994): 54–60; Ian Simmons. "Barber's Nightmare." *New Scientist* 140 (Oct. 9, 1993): 53; "Unruly Hair: No Fairy Tale." *Science News* 144 (Sept. 11, 1993): 164.

268 Anthony H. Rose. "Microbiological Production of Food and Drink." *Scientific American* 245 (Sept. 1981): 127–138.

269 James H. Aubert, Andrew M. Kraynik, and Peter B. Rand. "Aqueous Foams." *Scientific American* 254 (May 1986): 74–82.

270 Sharon Bertsch McGrayne. *Nobel Prize Women in Science: Their Lives, Struggles, and Momentous Discoveries.* New York: Carol Publishing, 1993; Lionel Milgrom. "The Assault on B-12." *New Scientist* 139 (Sept. 11, 1993): 39–44.

271 Harvey B. Hopps. "From the Sun, to the Moon and Beyond." *Chemtech* 23 (Feb. 1993): 47–51.

272 T. P. Coultate. *Food: The Chemistry of Its Components.* London: Royal Society of Chemists, 1989.

273 "Blue Roses?" *Science News* 149 (Jan. 20. 1995): 41.

274 C. Somerville. "Targeting of the Polyhydroxybutyrate Biosynthetic Pathway to the Plastids of *Arabidopsis thaliana* Results in High Levels of Polymer Accumulation." *Proceedings of the National Academy of Sciences USA* 91 (Dec. 20, 1994): 12760–12764.

275 Stu Borman. "Bacteria That Flourish Above 100 Degrees C Could Benefit Industrial Processing." *Chemical and Engineering News* 69 (Nov. 4, 1991): 31–34; Douglas Clark and Robert Kelly. "Hot Bacteria." *Chemtech* 20 (Nov. 1990): 654ff.

276 See 275.

277 Pete Moore. "Why Fires Go to Blazes." *New Scientist* 146 (June 3, 1995): 26–30.

278 Robert F. Service. "Ceramic Shrinks When the Heat Goes On." *Science* 272 (April 5, 1996): 30.

279 Esther Pohl Lovejoy. *Women Doctors of the World.* New York: Macmillan, 1957.

280 Debora MacKenzie. "Cheesy Feet Attract the First Bite." *New Scientist* 148 (Nov. 4, 1995): 7.

281 Knut Schmidt-Nielsen. "How Are Control Systems Controlled?" *American Scientist* 82 (Jan./Feb. 1994): 38–44.

282 "CFC Smuggling Threatens Ozone Recovery." *Science News* 149 (May 25, 1996): 331.

Astronomy

283 J. Kelly Beatty. "Instant Science on the Internet." *Sky & Telescope* 88 (Oct. 1994): 21; Paul Weissman. "Astronomers Feast on Big Bangs." *Physics World* 7 (Aug. 1994): 7.

284 Dan Durda. "Two by Two They Came." *Astronomy* 23 (Jan. 1995): 31–35; R. M. Hough et al. "Diamond and Silicon Carbide in Impact Melt Rock from the Ries Impact Crater." *Nature* 378 (Nov. 2, 1995): 41–44; Christian Koeberl. "Diamonds Everywhere." *Nature* 378 (Nov. 2, 1995): 17–18; Douglas Palmer. "Meteorite Showered Town with Diamonds." *New Scientist* 148 (Nov. 4, 1995): 18.

285 John Gribbin. "At the Third Pulsar the Time Will Be . . ." *New Scientist* 148 (Nov. 4, 1995): 11; Demetrios Matsakis, private communication.

286 Matthew L. Wald. "50 Billion Might As Well Be Infinity." *New York Times* Week in Review, Jan. 21, 1996, p. 2; John Noble Wilford. "Suddenly, Universe Gains 40 Billion More Galaxies." *New York Times,* Jan. 16, 1996, A1.

287 William J. Broad. "Meteoroids Hit Atmosphere in Atomic-Size Blasts." *New York Times,* Jan. 25, 1994, C5.

288 George Djorgovski, private communication.

289 Marcus Chown. "The Lab Between the Stars." *New Scientist* 148 (Dec. 16, 1995): 44–45; Marcus Chown. *Afterglow of Creation.* London: University Science Books, 1996; Nibel Henbest. "Hot Bubbles in Space." *New Scientist* 142 (April 30, 1994): 28–31; Stephen Taylor and David Williams. *Chemistry in Britain* 21 (Aug. 1993): 680–683.

290 David H. Hathaway. "Journey to the Heart of the Sun." *Astronomy* 23 (Jan. 1995): 38–43.

291 Victor E. Thoren. *The Lord of Uraniborg: A Biography of Tycho Brahe.* Cambridge, England: Cambridge University Press, 1990.

292 Keith P. Bowen. "Aging Eyeballs." *Sky & Telescope* 82 (Sept. 1991): 254–257.

293 Alan P. Boss. "Companions to Young Stars." *Scientific American* 273 (Oct. 1995): 38–43.

294 Brian Hayes. "Scanning the Heavens." *American Scientist* 82 (Nov./Dec. 1994): 512–516.

295 Mark A. Gordon. "VLBA—A Continent-Size Radio Telescope." *Sky & Telescope* 69 (June 1985): 487–490; Kenneth I. Kellermann and A. Richard Thompson. "The Very-Long-Baseline Array." *Scientific American* 258 (Jan. 1988): 54–63.

296 Keay Davidson. "Strangest Telescopes in the World." *New Scientist* 141 (Feb. 26, 1994):35–39; Leif J. Robinson. "Ice Fishing for Neutrinos." *Sky & Telescope* 88 (July 1994): 44–48; James S. Sweitzer. "The Last Observatory on Earth." *Mercury* 22 (Sept./Oct. 1993): 13–15.

297 See 296.

298 James B. Kaler. "Giants in the Sky: The Fate of the Sun." *Mercury* 22 (March/April 1993): 34–41.

299 See 298.

300 Jonathan I. Lunine. "Does Titan Have Oceans?" *American Scientist* 82 (March/April 1994): 134–143.

301 Marcus Chown. "The Lab Between the Stars." *New Scientist* 148 (Dec. 16, 1995): 44–45; Marcus Chown. *Afterglow of Creation.* London: Arrow, 1995; Jeff Hecht. "'Molecule of Life' Is Found in Space." *New Scientist* 142 (June 11, 1994): 4.

302 Janet G. Luhmann, James B. Pollack, and Lawrence Colin. "The Pioneer Mission to Venus." *Scientific American* 270 (April 1994): 90–97; Gerald Schubert and Curt Voey. "The Atmosphere of Venus." *Scientific American* 245 (July 1981): 66–74.

303 William J. Broad. "Space Probe Is Set to Visit Eros, an Earth-Grazing Asteroid." *New York Times,* Feb. 6, 1996, B5; Tom Gehrels. "Collisions with Comets and Asteroids." *Scientific American* 274 (March 1996): 54–61; John S. Lewis. *Rain of Iron and Ice.* New York: Addison-Wesley, 1996; S. J. Ostro et al. "Extreme Elongation of Asteroid 1620 Geographos from Radar Images." *Nature* 375 (June 8, 1995): 474–476.

304 Bradley E. Schaefer. "The Astronomical Sherlock Holmes." *Mercury* 22 (Jan./Feb. 1993): 9–13. Reprinted from the *Journal of the British Astronomical Association.*

305 France Allard. "A Very Cool Customer." *Nature* 378 (Nov. 30, 1995): 441–442; Jeff Hecht. "Shy Brown Dwarf Turns Up in Lepus." *New Scientist* 148 (Dec. 9, 1995): 19; T. Henry. "Brown Dwarf Reveals—At Last!" *Sky & Telescope* 91 (April 1996): 24–28; T. Nakajima et al. "Discovery of a Cool Brown Dwarf." *Nature* 378 (Nov. 30, 1995): 463–465.

306 L. Pearce Williams. "Maybe the Europeans Did See the Crab." *Mercury* 24 (Jan./Feb. 1995): 28; L. Pearce Williams. "The Supernova of 1054: A Medieval Mystery." In *The Analytic Spirit,* ed. Harry Woolf. Ithaca, N.Y.: Cornell University Press, 1981, pp. 328–347.

307 R. Cowen. "Ill-Fated Milky Way Neighbor Found." *Science News* 145 (April 9, 1994): 228.

308 R. Cowen. "One-Man Band: X-ray Source Plays Two Tunes." *Science News* 149 (March 9, 1996): 148–149; C. Kouveliotou et al. "A New Type of Transient High-Energy Source in the Direction of the Galactic Centre." *Nature* 379 (Feb. 29, 1996): 779–801; Christopher Thompson. "A New Rapid X-ray Repeater." *Nature* 379 (Feb. 29, 1996): 775–776.

309 R. Cowen. "Finding Planets Around Ordinary Stars." *Science News* 148 (Oct. 21, 1995): 260.

310 Richard Monastersky. "Large Meteorite Scar Identified in Virginia." *Science News* 146 (Aug. 20, 1994): 116–117; Richard A. Kerr. "Chesapeake Bay Impact Crater Confirmed." *Science* 269 (Sept. 22, 1995): 1672; Wylie Poag. "Meteoroid Mayhem in Ole Virginny." *Geology* 22 (August 1994): 691–696.

311 R. C. Balling Jr. and R. S. Cerveny. "Influence of Lunar Phase on Daily Global Temperatures." *Science* 267 (March 10, 1995): 1481–1483.

Physics

312 Richard I. Epstein et al. "Observation of Laser-induced Fluorescent Cooling of a Solid." *Nature* 377 (Oct. 12, 1995): 500–503.

313 Robert Service. "The Sandman Speaks." *Science* 264 (April 8, 1994): 200–201.

314 Ivars Peterson. "Trickling Sand: How an Hourglass Ticks." *Science News* 144 (Sept. 11, 1993): 167; X-l. Wu et al. "Why Hour Glasses Tick." *Physical Review Letters* 71 (Aug. 30, 1993): 1363–1366.

315 "New Light in Forests: Sylvanshine." *New York Times,* Sept. 13, 1994, B8.

316 C. Richard Taylor. "Freeloading Women." *Nature* 375 (May 4, 1995): 17; N. C. Heglund, P. A. Willems, M. Penta, and G. A. Cavagna. "Energy-Saving Gait Mechanics with Head-Supported Loads." *Nature* 375 (May 4, 1995): 52–54.

317 Brian Hayes. "Trails in the Trackless Sea." *American Scientist* 81 (Jan./Feb. 1993): 19–20.

318 John C. Salzsieder. "Exposing the Bathtub Coriolis Myth." *The Physics Teacher* 107 (Feb. 1994): 107.

319 David G. Grier. "On the Points of Melting." *Nature* 379 (Feb. 29, 1996): 773–774; Ivars Peterson. "Fine Points of Melting in Plasma Crystals." *Science News* 149 (March 9, 1996): 150; Alexander Piel. "Plasma Crystals Show New Order." *Physics World* 7 (Oct. 1994): 23–24.

320 James D. White. "The Role of Surface Melting in Ice Skating." *The Physics Teacher* 30 (Nov. 1992): 495–497; Samuel C. Colbeck. "Pressure Melting and Ice Skating." *American Journal of Physics* 63 (Oct. 1995): 888–890; "Skating on Thin Water." *Science News* 148 (Oct. 21, 1995): 268.

321 Carl W. Peterson. "The Fluid Physics of Parachute Inflation." *Physics Today* 46 (Aug. 1993): 32–39; Carl W. Peterson. "High-Performance Parachutes." *Scientific American* 262 (May 1990): 108–115.

322 Thomas A. Herring. "The Global Positioning System." *Scientific American* 274 (Feb. 1996): 44–50; Jennifer Quellette. "GPS Industry Prepares for Boom." *The Industrial Physicist* 1 (Dec. 1995): 8–12.

323 See 322.

324 John Crangle and Mike Gibbs. "Units and Unity in Magnetism: A Call for Consistency." *Physics World* 7 (Nov. 94): 31–32.

325 William R. Gregg. "An Old Optics Demonstration—Redone More Safely." *The Physics Teacher* 31 (Jan. 1993): 40.

326 Arthur E. Hallerburg. "House Bill No. 246 Revisited." *Proceedings of the Indiana Academy of Science* 84 (1995): 374–399; Robert J. Whitaker. "The Legislation of the Value of Pi." *The Physics Teacher* 31 (April 1993): 212–213.

327 Ivars Peterson. "Glimpsing Glueballs in Collider Debris." *Science News* 149 (Jan. 6, 1996): 5; James Sexton, Alessandro Vac-

carino, and Donald Weingarten. "Numerical Evidence for the Observation of a Scalar Glueball." *Physical Review Letters* 75 (Dec. 18, 1995): 4563; Donald Weingarten. "Quarks by Computer." *Scientific American* 274 (Feb. 1996): 104–108.

328 Marcus Chown. "Meandering Rivers Keep Themselves in Check." *New Scientist* 148 (Dec. 9, 1995): 19; T. B. Liverpool and S. F. Edwards. "The Dynamics of a Meandering River." *Physical Review Letters* 75 (1995): 3016.

329 Julian White. "Green Lights." *Physics World* 7 (Oct. 1994): 31–35.

330 Ivars Peterson. "Magnetism in Atomic Clusters." *Science News* 145 (May 7, 1994): 303.

331 L. Pearce Williams. *Michael Faraday.* New York: Basic Books, 1965; Henry A. Boorse and Lloyd Motz. *The World of the Atom.* New York: Basic Books, 1966.

332 Sam Austin and George Bertsch. "Halo Nuclei." *Scientific American* 272 (June 1995): 90–96.

333 William Grimes. "New Science Can Solve the Puzzle of Old Statues." *New York Times,* Nov. 1, 1995, B1.

334 John Clarke. "SQUIDS." *Scientific American* 271 (Aug. 1994): 46–53.

335 Thomas J. Clark, private communication, March 14, 1996; Marvin L. Cohen. "Harder Than Diamonds?" *The Sciences* 34 (May/June 1994): 26–30.

336 William L. Kerr and David S. Reid. "Thermodynamics and Frozen Foods." *The Physics Teacher* 31 (Jan. 1993): 52–55.

337 David Wilson. *Rutherford: Simple Genius.* Cambridge: MIT Press, 1983.

338 Edwin Kashy and Sharon McGrayne. "Electricity and Magnetism." *Encyclopaedia Britannica,* vol. 18. Chicago: Encyclopaedia Britannica, Inc., 1991 and later.

Mathematics and Computers

339 Jim Collins and Ian Stewart. "The Mathematical Springs in Insect Steps." *New Scientist* 143 (Oct. 8, 1994): 36–40.

340 Ivars Peterson. "Party Numbers." *Science News* 144 (July 17, 1993): 46–47.

341 See 340.

342 Natalie S. Glance and Bernardo A. Huberman. "The Dynamics of Social Dilemmas." *Scientific American* 270 (March 1994): 76–81.

343 Brian Hayes. "Waiting for 01-01-00." *American Scientist* 83 (Jan./Feb. 1995): 12–15; Ivars Peterson. "Reviving Software Dinosaurs." *Science News* 144 (Aug 7, 1993): 88–89.

344 Ivars Peterson. *Fatal Defect: Chasing Killer Computer Bugs*. New York: Times Books, 1995; Peter G. Neumann. *Computer-Related Risks*. Reading, Mass.: Addison-Wesley, 1995.

345 Mitchel Resnick. "Changing the Centralized Mind." *Technology Review* 97 (July 1994): 32–40; Mitchel Resnick. *Turtles, Termites, and Traffic Jams: Explorations in Massively Parallel Microworlds*. Cambridge: MIT Press, 1994.

346 Philip Yam. "Branching Out." *Scientific American* 271 (Nov. 1994): 26–30.

347 See 346.

348 S. Chandrasekhar. *Newton's Principia for the Commoner*. New York: Oxford University Press, 1995; I. Bernard Cohen and Richard S. Westfall, eds. *Newton*. London: Norton, 1995; David Hughes. "On the Shoulders of Giants." *Nature* 376 (Aug. 3, 1995): 395; John Maddox. "Is the *Principia* Publishable Now?" *Nature* 376 (Aug. 3, 1995): 385.

349 Masakazu Konishi. "Listening with Two Ears." *Scientific American* 268 (April 1993): 66–73.

350 "Computers Call Sand People Back from the Dead." *New Scientist* 115 (July 2, 1987): 38.

351 Douglas M. Campbell. "Beauty and the Beast: The Strange Case of André Bloch." *The Mathematical Intelligencer* 7, no. 4 (1985): 36–38; Henri Cartan and Jacqueline Ferrand. "The Case of André Bloch." *The Mathematical Intelligencer* 10, no. 1 (1988): 23–26; D. Huylebrouck. "Captain Mangin-Bocquet's Contribution to Mathematics." *The Mathematical Intelligencer* 16, no. 1 (1994): 8–9.

352 W. Wayt Gibbs. "Software's Chronic Crisis." *Scientific American* 271 (Sept. 1994): 86–95.

353 See 352.

354 Brian Hayes. "The Magic Words Are Squeamish Ossifrage." *American Scientist* 82 (July/Aug. 1994): 312–316; Gina Kolata. "The Assault on . . ." *New York Times,* March 22, 1994, C1ff; Gary Taubes. "Small Army of Code-Breakers Conquers a 129-Digit Giant" and "How to Make a Prime Cut." *Science* 264 (May 6, 1994): 776–777.

355 See 354.

356 Gerard J. Holzmann and Björn Pehrson. "The First Data Networks." *Scientific American* 270 (Jan. 1994): 124–129; Gerard J. Holzmann and Björn Pehrson. *The Early History of Data Networks.* Los Alamitos, Calif.: IEEE Computer Society Press, 1995.

357 Leigh Dayton. "Neural Nets Unearth Secrets in Stone." *New Scientist* 142 (April 16, 1994): 21; Robert Matthews and Tom Merriam. "A Bard by Any Other Name." *New Scientist* 141 (Jan. 22, 1994): 23–27; Gary Stix. "Bad Apple Picker." *Scientific American* 271 (Dec. 1994): 44–45; Paul Wallich. "Digital Dyslexia." *Scientific American* 265 (Oct. 1991): 36.

358 I am indebted to Benno Artmann, Technical University of Darmstadt (Germany).

359 William Menasco and Lee Rudolph. "How Hard Is It to Untie a Knot?" *American Scientist* 83 (Jan./Feb. 1995): 38–49; Carlo Rovelli. "Knot Theory and Space-Time." *Science and the Future 1993.* Chicago: Encyclopaedia Britannica, Inc., 1994.

360 See 359.

361 Morris Kline. *Mathematics in Western Culture*. New York: Oxford University Press, 1953.

362 John Horgan. "The Death of Proof." *Scientific American* 269 (Oct. 1993): 90–103.

363 Robert R. Birge. "Protein-Based Computers." *Scientific American* 272 (March 1995): 90–95; Robert R. Birge. "Protein-Based Three-Dimensional Memory." *American Scientist* 82 (July/Aug. 1994): 348–355.

364 Benno Artmann, Technical University of Darmstadt.

365 Steven J. Brams and Alan D. Taylor. "An Envy-free Cake Division Protocol." *The American Mathematical Monthly* 102 (Jan. 1995): 9–19; Will Hively. "Dividing the Spoils." *Discover* 16 (March 1995): 49–57; Alan Taylor and Steven J. Brams. *Fair Division: From Cake-Cutting to Dispute Resolution*. Cambridge, England: Cambridge University Press, 1996.

▶ Index

About the Author

Sharon Bertsch McGrayne is the author of *365 Surprising Scientific Facts, Breakthroughs, and Discoveries* (Wiley, 1994), a light digest of recent scientific research. Her book *Nobel Prize Women in Science: Their Lives, Struggles, and Momentous Discoveries* (Carol, 1993) is a collection of biographies about 14 women scientists who either won a Nobel Prize or contributed substantially to a Nobel Prize won by someone else. McGrayne has been a frequent lecturer about women in science at colleges, universities, research institutions, and corporations here and abroad. A former newspaper reporter, she has co-authored and edited articles about physics for the *Encyclopaedia Britannica*. She is a graduate of Swarthmore College and lives in Seattle.

LaVergne, TN USA
07 January 2010
169080LV00004B/4/A